N. N. Yanenko

The Method of Fractional Steps

The Solution of Problems of Mathematical Physics in Several Variables

English Translation Edited by M. Holt

With 15 Figures

Springer-Verlag New York Heidelberg Berlin 1971

Professor N. N. Yanenko
U.S.S.R. Academy of Sciences, Siberian Branch
Computing Center, Novosibirsk, U.S.S.R.

Professor Maurice Holt
University of California, Berkeley, Cal., U.S.A.

Translated into English by Northrop Corporate Laboratories,
Hawthorne, Cal., U.S.A.

ISBN 0-387-05272-0 Springer-Verlag New York Heidelberg Berlin
ISBN 3-540-05272-0 Springer-Verlag Berlin Heidelberg New York

Title of the Russian Original Edition:
Metod drobnykh shagov resheniya mnogomernykh zadach matematicheskoi fiziki
Publisher: "Nauka", Sibirskoe Otdelenie, Novosibirsk 1967

Title of the French Translation:
Méthode à pas fractionnaires, résolutions de problèmes polydimensionnels de physique mathématique. Librairie Armand Colin, Paris 1968

Title of the German Translation:
Die Zwischenschrittmethode zur Lösung mehrdimensionaler Probleme der mathematischen Physik (Lecture Notes in Mathematics, Vol. 91), 1969
ISBN 0-387-04610-0 Springer-Verlag New York Heidelberg Berlin
ISBN 3-540-04610-0 Springer-Verlag Berlin Heidelberg New York

 Library of Congress Catalog Card Number 78-139953

Editor's Preface

The method of fractional steps, known familiarly as the method of splitting, is a remarkable technique, developed by N. N. Yanenko and his collaborators, for solving problems in theoretical mechanics numerically. It is applicable especially to potential problems, problems of elasticity and problems of fluid dynamics. Most of the applications at the present time have been to incompressible flow with free boundaries and to viscous flow at low speeds. The method offers a powerful means of solving the Navier-Stokes equations and the results produced so far cover a range of Reynolds numbers far greater than that attained in earlier methods. Further development of the method should lead to complete numerical solutions of many of the boundary layer and wake problems which at present defy satisfactory treatment.

As noted by the author very few applications of the method have yet been made to problems in solid mechanics and prospects for answers both in this field and other areas such as heat transfer are encouraging. As the method is perfected it is likely to supplant traditional relaxation methods and finite element methods, especially with the increase in capability of large scale computers.

The literal translation was carried out by T. Cheron with financial support of the Northrop Corporation. The editing of the translation was undertaken in collaboration with N. N. Yanenko and it is a pleasure to acknowledge his patient help and advice in this project. The edited manuscript was typed, for the most part, by Mrs. Arlene Martin while many graduate students in this Division assisted in setting and checking equations. The author and editor wish to thank all these helpers and express their appreciation to the staff of Springer-Verlag for the printing and arrangement of the translation.

Berkeley, November 1970

Maurice Holt

Preface to the English Edition

The author is happy that his book is now available to English-speaking mathematicians. In the three years since the publication of the original Russian edition the method of fractional steps has been rapidly developed. Unfortunately, the author was unable to introduce in the English edition of the book any essential changes which would reflect the present development of the method. He has simply given some bibliographical supplements with brief comments.

Works published recently and those which escaped the author's attention earlier have only confirmed him in the opinion that the method of fractional steps is in fact a necessary stage in the development of computational and theoretical mathematics. Evidence for this is to be found in the remarkable similarity of concepts in various branches of mathematics: the theory of semigroups, the notion of weak approximation and the correctness of the Cauchy problem for differential equations, the decomposition principle for differential inequalities and linear programming, splitting schemes in gas dynamics and the PIC method, factorized schemes in mathematical physics and economical analog devices. Such interconnection between seemingly heterogeneous subjects provides evidence not only of the unity of mathematics but also of the fundamental fact that the construction of algorithms and their optimization are now the most important objects of mathematics.

The author is deeply grateful to Prof. M. Holt for the initiative taken in this edition and for the great attention he paid to the delicate problem of translating new and, sometimes, rather exotic Russian mathematical terms. He expresses his gratitude to Y. I. Shokin and A. N. Valiullin for valuable help in the checking of the manuscript.

Novosibirsk, November 1970

N. N. Yanenko

Preface to the Russian Edition

The subject of this book is the method of fractional steps; it was first conceived several years ago and has since been developed rapidly as a means of constructing economical difference schemes.

The method of fractional steps enables us to solve problems which could not be treated by means of ordinary difference schemes—schemes of simple approximation, where at each step conditions of stability and consistency must be satisfied. This leads to a great simplicity of formulae; however, the schemes become less flexible, contain a smaller number of arbitrary parameters and, by the same token, cannot satisfy all the requirements which are imposed. On the other hand, in the scheme of fractional steps the transition from one stage of the calculation to the next is divided into a series of intermediate steps and it is not required to satisfy conditions of consistency with the original equations and stability criteria at each stage. As a result, this method allows for a choice of parameters which makes possible the construction of economical and exact schemes.

The method of fractional steps appears as the answer to a real need arising in numerical calculus—to create simple economical schemes for the solution of complicated problems of mathematical physics in several variables.

Beginning with the basic work of Peaceman, Rachford and Douglas (1955), the method was extended and improved in the works of American and Soviet authors: Douglas, Rachford, Baker, Oliphant, K. A. Bagrinovskii, S. K. Godunov, E. G. D'yakonov, G. I. Marchuk, A. A. Samarskii, and others.

At the present time the method of fractional steps is an essential element in the construction of schemes for solving complicated problems of mathematical physics in several variables. In spite of the fact that the method of fractional steps can still be further developed and has not yet received full theoretical justification, it is already used, not only as a means of construction of optimal algorithms, but also as an instrument of theoretical investigation of difference and differential equations.

The book is based on a course of lectures given by the author in the State Universities of the Urals, Novosibirsk, Tomsk and Kazakhstan from 1959 to the present time. The book is an attempt to interpret and, if possible, to give a unified presentation of the many schemes of fractional steps, which otherwise can only be found in journals and proceedings.

The author has tried to take into account, as far as possible, all works contributing to the development of the method of fractional steps but, obviously, the interpretation of the subject is not uniform and reflects the personal tastes of the author. In particular, the book does not deal with methods of *a priori* accuracy evaluation. The author uses the simpler methods of harmonic analysis of convergence. Also, the theory of local convergence criteria is given only cursory consideration. The author has concentrated most of his attention on methods of construction of effective schemes.

The book reflects equally the personal results obtained by the author and those of his collaborators. The author is particularly pleased to recall the joint work carried out with teams of young mathematicians, who studied the first implicit splitting schemes for the solutions of various problems of mathematical physics, and used them successfully in numerical applications. Particular mention should be made of N. N. Anuchina, V. A. Enal'skii, A. S. Zharikov, A. I. Zuev, A. N. Konovalov, V. E. Neuvazhaev, Yu. Ya. Pogodin, V. A. Suchkov, V. D. Frolov.

In 1962 there was a discussion in which E. G. D'yakonov, A. A. Samarskii, and B. L. Rozhdestvenskii took part. This resulted in a more profound understanding of the role of boundary conditions in the evaluation of the accuracy of schemes of fractional steps, and gave the impetus to the production of a whole series of works.

Joint work and frequent discussions with G. I. Marchuk and the following research workers in the Siberian Branch of the Academy of Sciences of the U.S.S.R.: Yu. E. Boyarintsev, G. V. Demidov, V. P. Il'in, B. G. Kuznetsov, M. M. Lavrent'iev, V. V. Penenko, Yu. N. Vatolin, led to further applications of the method of fractional steps and its theoretical justification. Considerable help in the checking of the manuscript was given by A. N. Valiullin. To all these comrades the author expresses his gratitude.

The author hopes that the book will be useful for scientific workers concerned with solving problems of mechanics and physics in several variables, and also for students of advanced courses in the Universities, specializing in computational mathematics.

Novosibirsk, November 1965

N. N. Yanenko

Contents

Chapter 1. Uniform schemes

Chapter 2. Simple schemes in fractional steps for the integration of parabolic equations

Chapter 3. Application of the method of fractional steps to hyperbolic equations

Chapter 4. Application of the method of fractional steps to boundary value problems for Laplace's and Poisson's equations

Chapter 1

Uniform Schemes

1.1 The class of problems under investigation The Cauchy problem in Banach space

In this course of lectures we consider primarily the system of differential equations of the type

$$\frac{\partial u(x, t)}{\partial t} = L(D)\, u(x, t) + f(x, t), \tag{1.1.1}$$

where

$$u(x,t) = \{u_1(x_1,\dots,x_m,t),\, u_2(x_1,\dots,x_m,t),\dots,u_n(x_1,\dots,x_m,t)\},$$
$$f(x,t) = \{f_1(x_1,\dots,x_m,t),\, f_2(x_1,\dots,x_m,t),\dots,f_n(x_1,\dots,x_m,t)\}$$

are vector functions of the vector space variable $x = (x_1, \dots, x_m)$ and of time t; $L(D)$ is a matrix linear differential operator with variable coefficients, $D = \{D_i\}$, $D_i = \partial/\partial x_i$, $i = 1, \dots, m$.

Eq. (1.1.1), when written in suffix notation, has the form

$$\left.\begin{aligned} &\frac{\partial u_i(x_1, \dots, x_m, t)}{\partial t} = \sum_{j,\alpha} a_{ij\alpha_1\dots\alpha_m}(x_1, \dots, x_m, t)\, D_1^{\alpha_1} \dots \\ &\dots D_m^{\alpha_m} u_j(x_1, \dots, x_m, t) + f_i(x_1, \dots, x_m, t), \\ &0 \le \alpha_k \le q_k, \quad k = 1, \dots, m, \quad i, j = 1, \dots, n. \end{aligned}\right\} \tag{1.1.2}$$

For the system (1.1.1) we can pose the Cauchy problem for the strip

$$|x| < \infty, \quad 0 \le t \le T < \infty,$$

with initial conditions

$$u(x, 0) = u_0(x) \tag{1.1.3}$$

or the mixed Cauchy problem in a cylinder

$$\Omega = G \times H,$$

where G is some region within the hyperplane $t = 0$ with a boundary γ; $H = \{0 \le t \le T\}$. In the second case to the initial data

$$u(x, 0) = u_0(x), \quad x \in G, \tag{1.1.4}$$

certain boundary conditions should be added, which are satisfied on the lateral surface $\Gamma = \gamma \times H$ of the region Ω

$$l(D)\, u = \varphi(x, t). \tag{1.1.5}$$

Here $l(D)$ is some differential operator which depends on symbols $D_0 = \partial/\partial t$; $D_1, \ldots, D_m$; and $\varphi(x, t)$ is the vector function defined on Γ.

Below, only the Cauchy problem in a strip is considered, and the mixed Cauchy problem is mentioned only in specific examples. This eliminates the need for analysis of boundary conditions. In the majority of cases the Cauchy problem under consideration possesses a periodic property; i.e., the coefficients, the second term on the right $f(x, t)$, the initial data, and in consequence, the solutions, are periodic functions of x.

We shall consider the more general Cauchy problem when initial data are given at the time t_1, $0 \le t_1 \le \mathrm{T}$, and it is required to find the solution of $u(x, t)$ of the system (1.1.1), for all t, $t_1 \le t \le \mathrm{T}$, which tends continuously to the function of the initial data $u(x, t_1)$ as $t \to t_1$.

It is assumed that the Cauchy problem has a unique solution for all instants of time t_1, $0 \le t_1 \le \mathrm{T}$ and that all derivatives of the solution $u(x, t)$ appearing in Eq. (1.1.2) are continuous provided that the function of the initial data $u(x, t_1)$ is a sufficiently smooth function of x. Such a solution is called classical. Denote the function $u(x, t)$ by $u(t)$, consider t as the parameter, and the function $u(x, t)$ as the element of a set of functions of x at fixed t. In this sense the one-parameter set of elements $u(t)$ corresponds to the solution $u(x, t)$ of the system (1.1.1). The operator $L(D)$ and the function $f(x, t)$ in Eq. (1.1.1) are denoted by $L(t, D)$ and $f(t)$, respectively.

Let $u(t)$ be the classical solution of the homogeneous Cauchy problem ($f = 0$) with initial data $u(t_1)$. Due to the assumption mentioned above, the relation

$$u(t_2) = S(t_2, t_1)\, u(t_1), \qquad 0 \le t_1 \le t_2 \le \mathrm{T}, \tag{1.1.6}$$

for a sufficiently smooth function $u(t_1)$ determines the linear transfer operator $S(t_2, t_1)$.

Let us assume that there exists a Banach space B of functions of x in which some set of smooth functions forms a dense class, and that the relation (1.1.6) could be applied to the functions of B. According to a known theorem concerning the expansion of operators this is possible, provided the operator $S(t_2, t_1)$ is bounded on smooth functions. All evaluations of operators are given in norms of the space B.

Definition. *The problems* (1.1.1) *and* (1.1.3) *are well posed if*

$$\| S(t, 0) \| \le M(\mathrm{T}), \qquad 0 \le t < \mathrm{T}. \tag{1.1.7}$$

The system (1.1.1) *is correct if*

$$\| S(t_2, t_1) \| \leq M(\mathrm{T}), \quad 0 \leq t_1 \leq t_2 \leq \mathrm{T}. \tag{1.1.8}$$

The system (1.1.1) *is uniformly correct if*

$$\| S(t_2, t_1) \| \leq e^{\alpha (t_2 - t_1)} \tag{1.1.9}$$

for all t_1, t_2 *which satisfy the condition* $0 \leq t_1 \leq t_2 \leq \mathrm{T}$, *where* α *is a constant which depends only on* T.

It is clear that a uniformly correct system is correct and that, if the system is correct, the Cauchy problem is well posed.

Below, only uniformly correct systems are considered. The property of correctness provides the law of composition for the family of bounded operators $S(t_2, t_1)$

$$S(t_3, t_1) = S(t_3, t_2)\, S(t_2, t_1). \tag{1.1.10}$$

Eq. (1.1.10) represents the Huygens-Hadamard principle; namely, the successive solution of the Cauchy problem within the intervals $t_0, t_1; t_1, t_2; \ldots; t_{m-1}, t_m$ is equivalent to the solution of the Cauchy problem within the interval t_0, t_m.

It is also assumed that the condition of strong continuity of the operator $S(t_2 . t_1)$ is fulfilled

$$\| S(t + \tau, t)\, u_0 - u_0 \| \to 0, \quad \tau \to 0, \tag{1.1.11}$$

for arbitrary $u_0 \in B$ and t.

A set of operators $S(t_2, t_1)$ which satisfies conditions (1.1.9), (1.1.10), and (1.1.11) forms a semigroup[1]. The Cauchy problem in Banach space is considered in Ch. 10.

Below, we will use the following notation: C_p is a space of p-fold continuous differentiable functions with the norm

$$\| u \|_{C_p} = \max_{x \in G} \{ |u|, |u'|, \ldots, |u^{(p)}| \}, \tag{1.1.12}$$

L_p is a space of functions of which the p-th potence is integrable, with the norm

$$\| u \|_{L_p} = \sqrt[p]{\int_G |u|^p \, dx}. \tag{1.1.13}$$

In the special case $C = C_0$, C is the space of the continuous functions with the maximum norm

$$\| u \|_C = \max_{x \in G} |u|. \tag{1.1.14}$$

[1] See [2, 4] and [99] for the solution of the Cauchy problem and the theory of semigroups, and also [1] and [72] for information regarding the functional analysis.

1.2 Uniform schemes

Let

$$\frac{u^{n+1} - u^n}{\tau} = \Lambda_1 u^{n+1} + \Lambda_0 u^n + F^n \tag{1.2.1}$$

be a two-layer difference scheme which corresponds to system (1.1.1). Here

$$u^n(x) = \{u_i(x_1, \ldots, x_m, n\tau)\}, \qquad F^n(x) = \{F_i(x_1, \ldots, x_m, n\tau)\}$$

are vector functions; $\Lambda_0 = \Lambda_0(t, \tau, h, T)$, $\Lambda_1 = \Lambda_1(t, \tau, h, T)$ are matrix difference operators with variable coefficients,

$$T = \{T_\alpha\}, \qquad \alpha = -q_\alpha, -q_\alpha + 1, \ldots, q_\alpha.$$

The shift operators T_α are determined from formulas

$$\begin{aligned} T_i u(x_1, \ldots, x_m) &= u(x_1, \ldots, x_i + h_i, \ldots, x_m); \\ T_{-i} u(x_1, \ldots, x_m) &= u(x_1, \ldots, x_i - h_i, \ldots, x_m); \qquad T_{-i} = T_i^{-1}. \end{aligned} \tag{1.2.2}$$

Eq. (1.2.1) in suffix notation is

$$\begin{aligned} \frac{u_i^{n+1} - u_i^n}{\tau} = &\sum_{j,\beta} b_{ij\beta_1 \ldots \beta_m}(x, t, \tau, h)\, T_1^{\beta_1} \ldots T_m^{\beta_m} u_j^{n+1} + \\ &+ \sum_{j,\beta} c_{ij\beta_1 \ldots \beta_m} T_1^{\beta_1} \ldots T_m^{\beta_m} u_j^n + F_i^n, \end{aligned} \tag{1.2.3}$$

where the indices β_s $(s = 1, \ldots, m)$ can be positive or negative integers.

We introduce a series of concepts. The schemes of type (1.2.1) are called uniform because the approximation of (1.2.1) is of a uniform nature regardless of the point x, t. (The uniform scheme was discussed in the work of A. N. Tikhonov and A. A. Samarskii [93].) The scheme (1.2.1) is called explicit, if the operator $E - \tau\Lambda_1$ in the space of network functions is represented by a matrix with one diagonal, and is called implicit in the opposite case. The operator Λ is called singular if, for any smooth function f, we have

$$\|\Lambda f\| \geq \frac{A}{h^\alpha}, \qquad \alpha > 0, \qquad A > 0,$$

where A does not depend on $h = \max h_i$.

The operators Λ_0, Λ_1 are called finite if $|\beta_s| \leq Q$ where Q does not depend on τ and h. Note that the operator which is inverse to a finite operator is not, in general, a finite operator.

In practice the terms in Eq. (1.2.1) are evaluated at network points, and the notation of Eq. (1.2.3) therefore should be supplemented by expressions referred to network indices

$$\begin{aligned} \frac{u^{n+1}_{ik_1 \ldots k_m} - u^n_{ik_1 \ldots k_m}}{\tau} = &\sum_{j,\beta} b_{ij\beta_1 \ldots \beta_m} u^{n+1}_{jk_1+\beta_1 \ldots k_m+\beta_m} \\ &+ \sum_{j,\beta} c_{ij\beta_1 \ldots \beta_m} u^n_{jk_1+\beta_1 \ldots k_m+\beta_m} + F^n_{ik_1 \ldots k_m}, \end{aligned} \tag{1.2.4}$$

where the indices $k_1, \ldots, k_m$ fix the point

$$x_1 = k_1 h_1, \ldots, \qquad x_m = k_m h_m.$$

In a theoretical study we consider the operators Λ_1, Λ_0 to be acting in the same space as the operator L of (1.1.1), and the functions $u^n(x)$ to belong to B. Let us take as the initial condition for (1.2.1):

$$u^0(x) = u_0(x). \tag{1.2.5}$$

Let $u^n(x)$ be the solution of the homogeneous Cauchy problem (1.2.1) and (1.2.5), $(F^n = 0)$; then the equation

$$u^{n+1}(x) = C(\tau, h; t + \tau, t)\, u^n(x), \qquad t = n\tau, \tag{1.2.6}$$

determines the difference step operator $C(\tau, h; t+\tau, t)$; while the equation

$$u^n(x) = C(\tau, h; t_2, t_1)\, u^m(x), \qquad 0 \le t_1 = m\tau \le t_2 = n\tau \le \mathrm{T}, \tag{1.2.7}$$

determines the difference transfer operator $C(\tau, h; t_2, t_1)$.

We also write

$$\begin{aligned} C_{nm} &= C(\tau, h; t_2, t_1); \qquad t_2 = n\tau, \ t_1 = m\tau; \\ C_n &= C_{nn-1}. \end{aligned} \tag{1.2.7'}$$

Definition. *The Cauchy problem* (1.2.1) *and* (1.2.5) *is correct, if*

$$\begin{aligned} \| C(\tau, h; t, 0) \| &\le M(\mathrm{T}), \qquad 0 \le t = n\tau \le \mathrm{T}, \\ \tau^2 + h^2 &\le \tau_0^2, \qquad M(\mathrm{T}) < \infty, \end{aligned} \tag{1.2.8}$$

where τ_0 is some sufficiently small constant. The scheme (1.2.1) *is correct, if*

$$\begin{aligned} &\| C(\tau, h; t_2, t_1) \| \le M(\mathrm{T}), \\ 0 \le t_1 = m\tau \le t_2 &= n\tau \le \mathrm{T}, \qquad M(\mathrm{T}) < \infty, \end{aligned} \tag{1.2.9}$$

and is uniformly correct, if

$$\begin{aligned} &\| C(\tau, h; t_2, t_1) \| \le e^{\omega(t_2 - t_1)}, \\ 0 \le t_1 &= m\tau \le t_2 = n\tau \le \mathrm{T}, \end{aligned} \tag{1.2.10}$$

where ω is a constant which depends only on T. *The scheme* (1.2.1) *is stable if*

$$\| C(\tau, h; t_2, t_1) \| \le 1, \qquad 0 \le t_1 = m\tau \le t_2 = n\tau \le \mathrm{T}; \tag{1.2.11}$$

is asymptotically stable, if

$$\| C(\tau, h; t_2, t_1) \| \to 0, \qquad t_2 \to \infty; \tag{1.2.11'}$$

and is strongly stable if

$$\begin{aligned} \| C(\tau, h; t+\tau, t) \| &\le 1 - \varepsilon(\tau, h, \mathrm{T}), \qquad \varepsilon > 0, \\ \varepsilon \to 0 \quad &for \quad \tau, h \to 0. \end{aligned} \tag{1.2.12}$$

These definitions can easily be extended to apply to the case of variable step size

$$\tau_1 = t_1 - t_0, \quad \tau_2 = t_2 - t_1, \ldots, \quad \tau_n = t_n - t_{n-1}.$$

Definition. *The scheme* (1.2.1) *is consistent with Eq.* (1.1.1), *if*

$$\frac{\| [C(\tau, h; t+\tau, t) - S(t+\tau, t)]\, u(x, t) \|}{\tau} \to 0 \qquad (1.2.13)$$

as $\tau \to 0$ *uniformly with respect to* t, $0 \leq t \leq T$, *for smooth solutions* $u(x, t)$ *of the problems* (1.1.1) *and* (1.1.3).

Definition. *The solution* $u^n(x)$ *of the problem* (1.2.1), (1.2.5) *converges to the solution* $u(x, t)$ *of the problems* (1.1.1), (1.1.3), *if, putting* $t = n\tau$,

$$\| u^n(x) - u(x, n\tau) \| = \| [C(\tau, h; n\tau, 0) - S(n\tau, 0)]\, u_0 \| \to 0 \qquad (1.2.14)$$

as $\tau \to 0$ *uniformly in the interval* $0 \leq t \leq T$ *for arbitrary* $u_0 \in B$[1].

Definitions of correctness, consistency, and convergence require the statement of law of proceeding to the limit

$$h = h(\tau), \quad h \to 0, \quad \tau \to 0. \qquad (1.2.15)$$

These are applicable to expressions (1.2.8)–(1.2.14).

Definition. *The scheme* (1.2.1) *is unconditionally (absolutely) correct if it is correct for any law of proceeding to the limit* $\tau, h \to 0$, *i.e., the condition* (1.2.9) *is valid within the quarter circle*

$$\tau^2 + h^2 \leq \tau_0^2, \quad \tau > 0, \quad h > 0 \qquad (1.2.16)$$

for sufficiently small τ_0. *The scheme* (1.2.1) *is said to be conditionally stable in the opposite case, i.e., if the condition* (1.2.9) *is not fulfilled for an arbitrary law of proceeding to the limit.*

In the case of conditional correctness condition (1.2.9) is valid within some incomplete neighborhood of zero in the quarter circle (1.2.16), the boundary of which passes through the point (0, 0).

The distinction between stability and strong stability is defined in the same way as that between absolute and conditional uniform correctness.

Definition. *The scheme* (1.2.1) *is absolutely consistent with Eq.* (1.1.1) *if condition* (1.2.13) *is satisfied in the neighborhood* (1.2.16), *and is conditionally consistent in the opposite case.*

Absolute and conditional convergence are defined analogously.

---

[1] We have defined convergence for $t = n\tau$. But with the restriction on C_{nm}, convergence for $t = n\tau$ implies convergence for all t.

Convergence theorem[1]. *If* (i) *the difference and the differential Cauchy problems are correct*; (ii) *the operator* $\Lambda_1 + \Lambda_0$ *approximates* $L: \Lambda_1 + \Lambda_0 \sim \sim L$; (iii) $\|(E - \tau\Lambda_1)^{-1}\| \le N(T)$, *then the difference solution of the Cauchy problem* (1.2.1), (1.2.5) *converges to the solution of the differential Cauchy problem* (1.1.1), (1.1.3).

Proof. Let $u(x, t) \in C_q$ be the solution of (1.1.1), (1.1.3), corresponding to the initial condition $u_0(x) \in C_p$, and $u^n(x)$ be the solution of the difference problem (1.2.1), (1.2.5) with the same initial conditions.

The quantity

$$v^n(x) = u^n(x) - u(x, n\tau) \tag{1.2.17}$$

satisfies the difference equation

$$\frac{v^{n+1} - v^n}{\tau} = \Lambda_1 v^{n+1} + \Lambda_0 v^n + R_{n+1}, \tag{1.2.18}$$

where

$$R_{n+1} = -\left\{\frac{u[x, (n+1)\tau] - u(x, n\tau)}{\tau} - \right.$$
$$\left. - \Lambda_1 u[x, (n+1)\tau] - \Lambda_0 u(x, n\tau)\right\} \tag{1.2.19}$$

with initial data

$$v^0 = 0. \tag{1.2.20}$$

According to condition (ii) (approximation condition) we have

$$\max_n \|R_n\| \to 0 \tag{1.2.21}$$

as $\tau \to 0$. Using condition (iii) of the convergence theorem, Eq. (1.2.18) can be represented in the form:

$$v^{n+1} = C_{n+1} v^n + r_{n+1}, \tag{1.2.22}$$

$$C_{n+1} = (E - \tau\Lambda_1)^{-1}(E + \tau\Lambda_0), \quad r_{n+1} = \tau(E - \tau\Lambda_1)^{-1} R_{n+1}. \tag{1.2.23}$$

It is easy to verify the expression

$$v^n = C_{n0} v^0 + \sum_{\alpha=1}^{n} C_{n\alpha} r_\alpha, \tag{1.2.24}$$

where

$$C_{n\alpha} = C_n C_{n-1} \ldots C_{\alpha+1}. \tag{1.2.25}$$

It follows from the correctness condition that

$$\|v^n\| \le \|C_{n0}\| \, \|v^0\| + \sum_{\alpha=1}^{n} \|C_{n\alpha}\| \, \|r_\alpha\|$$
$$\le n\tau N M \max_{k=1,\ldots,n} \|R_k\| = t M N \max_{k=1,\ldots,n} \|R_k\| \tag{1.2.26}$$

---

[1] Analogous convergence conditions were obtained for the first time by V. S. Ryaben'kii [5] and N. N. Meiman [6]. For simplicity we consider only the case $f = F = 0$.

and from this we have

$$\|v^n\| \to 0, \quad \text{as} \quad \tau \to 0$$

uniformly in the interval $(0, T)$. This is the required proof. □

Equivalence theorem[1]. *If the scheme* (1.2.1) *is consistent with Eq.* (1.1.1), *then in order to make the solution* $u^n(x)$ *of the problem* (1.2.1), (1.2.5) *converge to the solution* $u(x, t)$ *of the correct problem* (1.1.1), (1.1.3) *in* B, *it is necessary and sufficient that the problem* (1.2.1), (1.2.5) *be correct.*

The correctness of the difference problem is an independent requirement which does not follow from the approximation properties. It is proved below in the case of an equation with constant coefficients.

Let

$$\frac{\partial u}{\partial t} = L(D)\, u \tag{1.2.27}$$

be an equation with constant coefficients, where u is a scalar function of $x_1, \ldots, x_m, t$;

$$L(D) = \sum_{\alpha} a_{\alpha_1 \ldots \alpha_m} D_1^{\alpha_1} \ldots D_m^{\alpha_m} \tag{1.2.28}$$

is a polynomial in the differential operators D_i.

A simple correctness criterion is satisfied in this case. Consider the harmonic function

$$u = u_0\, e^{\omega t + i k x}, \qquad u_0 = \text{const}, \tag{1.2.29}$$

where k is a vector with integer components and $k\,x$ is a scalar product. In order to make Eq. (1.2.29) the solution of Eq. (1.2.27) it is necessary and sufficient that ω and k be connected by the following relation (the so-called dispersion or characteristic equation)

$$\omega = L(i\,k). \tag{1.2.30}$$

The system (1.2.27) is correct in $L_2(-\pi, \pi)$ if, and only if

$$\mathrm{Re}\,\omega(k) \leq \mu_1 \tag{1.2.31}$$

for all k, where μ_1 is a constant which does not depend on k.

Let

$$\frac{u^{n+1} - u^n}{\tau} = L\left(\frac{\Delta}{h}\right) u^n \tag{1.2.32}$$

be an explicit uniform scheme consistent with Eq. (1.2.27); Δ/h is some approximation to D, for example,

$$\frac{\Delta_i}{h_i} = \frac{T_i - E}{h_i} \sim D_i.$$

---

[1] The equivalence theorem has been proved by P. D. Lax [3, 4] for homogeneous equations. It was extended to non-homogeneous problems by R. D. Richtmyer [7].

Then Eq. (1.2.29) may be the solution of Eq. (1.2.32), under the condition (difference dispersion equations)

$$\frac{e^{\omega\tau}-1}{\tau} = L\left(\frac{e^{ikh}-1}{h}\right)$$
$$= \sum_{\alpha} a_{\alpha_1 \ldots \alpha_m} \left(\frac{e^{ik_1h_1}-1}{h_1}\right)^{\alpha_1} \cdots \left(\frac{e^{ik_mh_m}-1}{h_m}\right)^{\alpha_m}. \qquad (1.2.33)$$

The necessary and sufficient correctness condition for the scheme (1.2.32) is the inequality

$$\mathrm{Re}\,\omega(\tau, h, k) \le \mu_2, \qquad (1.2.34)$$

where ω is determined from Eq. (1.2.33); μ_2 does not depend on k, τ, h. For a bounded value of $k \le K$

$$\omega(\tau, h, k) \to \omega(k) = L(i\,k) \quad \text{as} \quad \tau, h \to 0 \qquad (1.2.35)$$

and consequently, the dispersion equation (1.2.33) approximates Eq. (1.2.30) for the class of functions having a finite Fourier transform, and the difference scheme (1.2.32) will be correct for this class.

However, in practice the situation is different: τ and h are small but finite and k is as large as we please or sufficiently large[1]. It follows that $\omega(\tau, h, k)$ could differ arbitrarily from $\omega(k)$, and condition (1.2.34) would not follow from condition (1.2.31).

A harmonic stability analysis is very useful for equations with constant coefficients. As opposed to the ε-schemes, in which a perturbation of initial data at an individual network point is considered together with analysis of its further propagation in the phase space $(x_1, \ldots, x_m, t)$, in harmonic analysis we consider an initial perturbation of the harmonic type

$$\delta u = \delta u_0\, e^{i(k_1x_1+k_2x_2+\ldots+k_mx_m)}$$

and investigate its further propagation. The harmonic criterion of stability may be formulated in the following form: if the amplitude of each harmonic perturbation increment is not stronger than $e^{\mu t}$, then the scheme is correct. Perturbation of the type $\delta u_0\, e^{ik_1x_1}$ in the initial data or in the process of calculation (round-off error) is called the error with respect to x_1, a perturbation of the type $\delta u_0\, e^{ik_2x_2}$ is called the error with respect to x_2, etc.

Thus, the difference schemes should satisfy the following two independent requirements:

(i) consistency,

(ii) correctness.

---

[1] If the difference problem is considered in the same space B as the initial Cauchy problem, then k is as large as we please. If the difference scheme is considered in the space of a network function then $k \sim 1/h$.

As will be seen later, these requirements are not only independent but they also contradict each other to a certain extent.

In addition to requirements (i) and (ii), difference schemes should also satisfy a series of requirements which are less categorical but necessary in practice.

In the first place there is the economy of the scheme, in other words, the economy of the computer time required. Economy of difference schemes is not only a means of saving machine time, but in certain cases, is an essential requirement for carrying out the scheme in program form.

The criteria of consistency and correctness become complicated when applied to the integration of nonlinear partial differential equations. Therefore, it is almost mandatory that the scheme satisfy conservation conditions[1].

These requirements could be added to, but one fact is clear; namely, that the construction of a good difference scheme is a very complex problem.

1.3 Examples

We shall illustrate this introduction with certain examples. Let us consider four schemes for the integration of the heat conduction equation

$$\frac{\partial u}{\partial t} = a^2 \frac{\partial^2 u}{\partial x^2}, \qquad a = \text{const} \neq 0. \tag{1.3.1}$$

(i) *The "cross" scheme* (Richardson scheme):

$$\frac{u^{n+1} - u^{n-1}}{2\tau} = a^2 \frac{\Delta_1 \Delta_{-1}}{h^2} u^n, \quad \Delta_1 = T_1 - E, \quad \Delta_{-1} = E - T_{-1}. \tag{1.3.2}$$

It is easy to verify that this is of second order. Assume that $\varrho = e^{\omega\tau}$. Then the dispersion equation for scheme (1.3.2) is

$$\varrho^2 + 8r \sin^2 \frac{kh}{2} \varrho - 1 = 0; \qquad r = \frac{a^2 \tau}{h^2}. \tag{1.3.3}$$

It follows from this that

$$\varrho_{1,2} = -4r \sin^2 \frac{kh}{2} \pm \sqrt{\left(4r \sin^2 \frac{kh}{2}\right)^2 + 1}.$$

The norm of the step operator is $4r + \sqrt{1 + (4r)^2}$. For any τ, h, scheme (1.3.2) is absolutely unstable. Therefore, the "cross" scheme is absolutely consistent but is absolutely unstable.

---

[1] A difference scheme is called conservative if its conservation laws are satisfied identically.

(ii) *The "rhombus" scheme* (Dufort-Frankel scheme):

$$\frac{u^{n+1} - u^{n-1}}{2\tau} = \frac{a^2}{h^2}\left[T_1 u^n + T_{-1} u^n - (u^{n-1} + u^{n+1})\right]. \qquad (1.3.4)$$

The scheme can be reduced to the form

$$\frac{u^{n+1} - u^{n-1}}{2\tau} = \frac{a^2}{h^2}\Delta_1 \Delta_{-1} u^n - \frac{a^2 \tau^2}{h^2}\,\frac{u^{n+1} - 2u^n + u^{n-1}}{\tau^2}. \qquad (1.3.4')$$

The dispersion equation is then

$$\varrho^2 - \frac{4r\cos k\,h}{1+2r}\varrho + \frac{2r-1}{2r+1} = 0.$$

It follows from this that

$$\varrho = \frac{2r\cos k\,h \pm \sqrt{(2r\cos k\,h)^2 - 4r^2 + 1}}{2r+1} = \frac{2r\cos k\,h \pm \sqrt{1-\varepsilon}}{2r+1},$$

where

$$\varepsilon = 4r^2 \sin^2 k\,h \geq 0.$$

If $\varepsilon > 1$, then roots ϱ_1 and ϱ_2 are complex conjugates and they have the modulus

$$\sqrt{\frac{2r-1}{2r+1}} < 1.$$

If $\varepsilon \leq 1$, then $\sqrt{1-\varepsilon} = \vartheta \leq 1$,

$$|\varrho| = \frac{2r\cos k\,h \pm \vartheta}{2r+1} \leq 1.$$

Thus, the scheme is absolutely stable. Using Eq. (1.3.4′) we see that the "rhombus" scheme is consistent with the equation

$$\frac{\partial u}{\partial t} = a^2\frac{\partial^2 u}{\partial x^2} - \frac{a^2\tau^2}{h^2}\frac{\partial^2 u}{\partial t^2}.$$

With the following law of proceeding to the limit

$$\frac{a^2\tau}{h^2} = r = \text{const}$$

the "rhombus" scheme is consistent with the heat conduction equation (1.3.1). With the limit law

$$\frac{a\tau}{h} = \varkappa = \text{const}$$

the "rhombus" scheme is consistent with the equation of hyperbolic type

$$\frac{\partial u}{\partial t} = a^2\frac{\partial^2 u}{\partial x^2} - \varkappa^2\frac{\partial^2 u}{\partial t^2}. \qquad (1.3.5)$$

Consequently, the "rhombus" scheme is absolutely stable and explicit, and is non-absolutely consistent with the heat conduction equation.

(iii) *The explicit two-layer scheme:*

$$\frac{u^{n+1} - u^n}{\tau} = a^2\frac{\Delta_1\Delta_{-1}}{h^2}u^n. \qquad (1.3.6)$$

The dispersion equation is

$$\varrho = 1 - 4r \sin^2 \frac{k h}{2}. \tag{1.3.7}$$

The scheme (1.3.6) is correct if

$$r = \frac{a^2 \tau}{h^2} \leq \frac{1}{2}. \tag{1.3.8}$$

That is, it is conditionally stable. It is easy to see that scheme (1.3.6) is absolutely consistent with Eq. (1.3.1). This means that scheme (1.3.6) is absolutely consistent with Eq. (1.3.1) but is conditionally stable.

(iv) *The implicit two-layer scheme* (Crank-Nicholson scheme):

$$\frac{u^{n+1} - u^n}{\tau} = \frac{a^2}{h^2} \Delta_1 \Delta_{-1} [\alpha\, u^{n+1} + (1 - \alpha)\, u^n], \quad 0 \leq \alpha \leq 1. \tag{1.3.9}$$

The dispersion equation is

$$\varrho = \frac{1 - 4r(1 - \alpha) \sin^2 \frac{k h}{2}}{1 + 4r\,\alpha \sin^2 \frac{k h}{2}}. \tag{1.3.10}$$

Clearly, scheme (1.3.9) is absolutely consistent with Eq. (1.3.1). For $\alpha \geq 1/2$ scheme (1.3.9) is also absolutely stable in L_2. For $\alpha = 1$ scheme (1.3.9) is absolutely stable in C since it satisfies the maximum principle. For $0 \leq \alpha \leq 1/2$, $r \leq 1/[2(1 - 2\alpha)]$, scheme (1.3.9) is stable in L_2.

Only implicit schemes are absolutely stable and consistent in these examples. Evidently this is true in general cases.

Schemes which are at the same time absolutely consistent and absolutely correct are especially convenient for practical calculations.

1.4 The method of factorization (sweep)

To put the scheme (1.3.9) into effect we can apply the method of factorization, in which the operator of the second order $E - \alpha r \Delta_1 \Delta_{-1}$ is treated as the product of two operators of first order. For the difference equation of the second order

$$A_i u_{i-1} + B_i u_i + C_i u_{i+1} = f_i, \quad i = 1, \ldots, N, \tag{1.4.1}$$

the sweep formulas are

$$u_i = X_i u_{i+1} + Y_i; \tag{1.4.2a}$$

$$X_i = -\frac{C_i}{B_i + A_i X_{i-1}}, \quad Y_i = \frac{f_i - A_i Y_{i-1}}{B_i + A_i X_{i-1}}. \tag{1.4.2b}$$

Assuming that $u_i^{n+1} = u_i$ we have, for scheme (1.3.9),

$$A_i = C_i = -\alpha r; \quad B_i = 1 + 2\alpha r; \quad f_i = [E + (1 - \alpha) r \Delta_1 \Delta_{-1}]\, u_i^n. \tag{1.4.3}$$

It follows from this that

$$X_i = \frac{\alpha r}{(1+2\alpha r) - \alpha r X_{i-1}} = \frac{1}{\left(2+\frac{1}{\alpha r}\right) - X_{i-1}};$$

$$Y_i = \frac{f_i + \alpha r Y_{i-1}}{(1+2\alpha r) - \alpha r X_{i-}} = \frac{Y_{i-1} + \frac{1}{\alpha r} f_i}{\left(2+\frac{1}{\alpha r}\right) - X_{i-1}}. \tag{1.4.3'}$$

Quantities X_0 and Y_0 are determined from boundary conditions at the left end. For example, if the following boundary problem is solved for Eq. (1.3.9)

$$\text{(a)}\ \frac{\partial u(0,t)}{\partial x} = 0; \quad \text{(b)}\ u(1,t) = 1; \qquad u(x,0) = u_0(x), \tag{1.4.4}$$

then X_0 and Y_0 are determined from the sweep relation

$$u_0 = X_0 u_1 + Y_0. \tag{1.4.5}$$

In order to satisfy the boundary condition (1.4.4a) it is sufficient to assume that

$$X_0 = 1, \qquad Y_0 = 0. \tag{1.4.6}$$

Then, using relations (1.4.2), quantities X_i and Y_i are determined in succession. From the right-hand condition (1.4.4b) we determine u_{N+1} and by applying Eq. (1.4.2a) from right to left we determine u_i.

It is easy to see that the iteration scheme (1.4.2) and (1.4.3) is spatially stable; i.e., the errors do not increase during the successive calculation of X_i, Y_i and u_i. The iteration algorithm is very effective, the number of operations per step in scheme (1.3.9) being only 5 times larger than the number of operations in the simple scheme (1.3.6).

It is evident from this that the use of the implicit scheme is less expensive, provided that $\tau_2/\tau_1 > 5$, where τ_1 is the step of the explicit scheme, and τ_2 is the step of the implicit scheme.

1.5 The method of matrix factorization

A different approach is needed if we then wish to solve the heat conduction equation in several dimensions. Consider, for example, the two-dimensional heat conduction equation

$$\frac{\partial u}{\partial t} = a^2\left(\frac{\partial^2 u}{\partial x^2} + \frac{\partial^2 u}{\partial y^2}\right) \tag{1.5.1}$$

satisfied in the parallelepiped

$$0 \le x \le 1; \quad 0 \le y \le 1; \quad 0 \le t \le T,$$

and let us pose for this equation the first Cauchy boundary value problem

$$u(x, y, 0) = u_0(x, y), \qquad (x, y) \in G;$$
$$u(x, y, t) = g(x, y, t), \qquad (x, y) \in \gamma;$$

where $G = \{0 < x < 1, 0 < y < 1\}$; γ is the boundary of G.

By analogy with the one-dimensional case, we use the uniform implicit scheme

$$\frac{u^{n+1} - u^n}{\tau} = \Lambda[\alpha u^{n+1} + (1 - \alpha) u^n], \tag{1.5.2}$$

where

$$\Lambda = \Lambda_1 + \Lambda_2; \quad \Lambda_1 = a^2 \frac{\Delta_1 \Delta_{-1}}{h_1^2}; \quad \Lambda_2 = a^2 \frac{\Delta_2 \Delta_{-2}}{h_2^2}. \tag{1.5.3}$$

In this case the following system of equations should be solved at each step

$$-\alpha r_1 (u_{i-1j} + u_{i+1j}) - \alpha r_2 (u_{ij-1} + u_{ij+1}) + [1 + 2\alpha (r_1 + r_2)] u_{ij} = f_{ij}, \tag{1.5.4}$$

where

$$f_{ij} = [E + (1 - \alpha) \tau \Lambda] u_{ij}^n; \qquad u_{ij} = u_{ij}^{n+1}.$$

The method of matrix factorization is used to solve Eq. (1.5.4).

This method was developed by a group of Soviet scientists (M. V. Keldysh, I. M. Gel'fand, K. I. Babenko, O. V. Lokutsievskii, N. N. Chentsov and others) [8].

G. I. Marchuk [9] has successfully applied the method of vector and matrix factorization to the solution of problems of nuclear physics.

We shall quickly describe this method in the case of the two-dimensional heat conduction equation (1.5.1). Eq. (1.5.4) can be represented in the matrix form

$$A_i \boldsymbol{u}_{i-1} + B_i \boldsymbol{u}_i + C_i \boldsymbol{u}_{i+1} = \boldsymbol{f}_i, \tag{1.5.5}$$

where $\boldsymbol{u}_i, \boldsymbol{f}_i$ are vectors $\{u_{ij}\}$, $\{f_{ij}\}$; the matrixes A_i, B_i and C_i act in the N_2-dimensional space of vectors $\boldsymbol{u}_i$

$$A_i = C_i = \begin{Vmatrix} -\alpha r_1 & 0 & \dots & 0 \\ 0 & -\alpha r_1 & \dots & 0 \\ \dots & \dots & \dots & \dots \\ 0 & 0 & \dots & -\alpha r_1 \end{Vmatrix} = -\alpha r_1 I, \tag{1.5.6}$$

$$B_i = \begin{Vmatrix} 1 + 2\alpha(r_1 + r_2) & -\alpha r_2 & 0 & \dots & 0 \\ -\alpha r_2 & 1 + 2\alpha(r_1 + r_2) & -\alpha r_2 & \dots & 0 \\ \dots & \dots & \dots & \dots & \dots \\ 0 & 0 & 0 & \dots & 1 + 2\alpha(r_1 + r_2) \end{Vmatrix}. \tag{1.5.7}$$

The form of the matrices A_i and B_i corresponds to the first boundary value problem for a rectangle

$$x_i = i h_1, \quad i = 0, 1, \dots, N_1 + 1;$$
$$y_j = j h_2, \quad j = 0, 1, \dots, N_2 + 1;$$

indices 0, N_1+1, N_2+1 refer to the sides of the rectangle boundary. By analogy with the iteration method it is assumed that

$$\boldsymbol{u}_i = X_i \boldsymbol{u}_{i+1} + \boldsymbol{Y}_i, \tag{1.5.8}$$

where X_i represent matrixes, and $\boldsymbol{u}_i$, $\boldsymbol{Y}_i$ the vectors.

Substituting Eq. (1.5.8) into Eq. (1.5.5) we obtain

$$(B_i + A_i X_{i-1}) \boldsymbol{u}_i + C_i \boldsymbol{u}_{i+1} = \boldsymbol{f}_i - A_i \boldsymbol{Y}_{i-1}. \tag{1.5.9}$$

By multiplying the left-hand side of Eq. (1.5.9) by matrix $(B_i + A_i X_{i-1})^{-1}$ we find

$$\boldsymbol{u}_i = -(A_i X_{i-1} + B_i)^{-1} C_i \boldsymbol{u}_{i+1} + (A_i X_{i-1} + B_i)^{-1}(\boldsymbol{f}_i - A_i \boldsymbol{Y}_{i-1}). \tag{1.5.10}$$

Comparing Eqs. (1.5.10) and (1.5.8) we deduce the recurrence relation

$$\begin{aligned} X_i &= -(A_i X_{i-1} + B_i)^{-1} C_i, \\ \boldsymbol{Y}_i &= (A_i X_{i-1} + B_i)^{-1} (\boldsymbol{f}_i - A_i \boldsymbol{Y}_{i-1}). \end{aligned} \tag{1.5.11}$$

From the boundary conditions we find

$$\begin{aligned} X_0 = 0, \quad \boldsymbol{Y}_0 = \boldsymbol{u}_0, \quad \boldsymbol{u}_0 = \{g(0, y_j, t)\}, \\ 0 \le y_j \le 1, \quad y_j = j h_2. \end{aligned} \tag{1.5.12}$$

Conditions (1.5.12) provide the initial data for recurrence formulae (1.5.11) which enable us to determine X_i, $\boldsymbol{Y}_i$ in succession up to $i = N_1$. The relation

$$\boldsymbol{u}_{N_1} = X_{N_1} \boldsymbol{u}_{N_1+1} + \boldsymbol{Y}_{N_1}; \qquad \boldsymbol{u}_{N_1+1} = \{g(1, y_j, t)\}, \qquad 0 \le y_j \le 1, \tag{1.5.13}$$

expresses the vector $\boldsymbol{u}_{N_1}$ in terms of the known vector $\boldsymbol{u}_{N_1+1}$; then u_i is determined from relation (1.5.8). Thus, the scheme of matrix sweep is wholly identical with the usual method of sweep, except that we apply the sweep process, not to the scalar qualities X_i, Y_i, u_i but instead to vectors $\boldsymbol{Y}_i$, $\boldsymbol{u}_i$, and the matrix X_i; also, the coefficients A_i, B_i and C_i become matrices and all operations are considered as operations with matrices and vectors.

If the coefficient of the heat conduction equation is variable, then for each i we need to invert a matrix of order N_2, and this is a difficult operation. Therefore, the use of matrix factorization for the heat conduction equation in a rectangular region is recommended only if N_2 is not too large. A still more complicated algorithm is needed for the three-dimensional problem. One reason for the sharp increase in the number of operations is the increase in dimension of the difference operator at the upper step, as compared with the one-dimensional case. We can try to reduce the dimension of this operator. For example, we can use the following approximation, instead of the difference scheme (1.5.2)

$$\frac{u^{n+1} - u^n}{\tau} = \Lambda_1 u^{n+1} + \Lambda_2 u^n. \tag{1.5.14}$$

Then the solution of the implicit scheme (1.5.14) will be reduced to the usual sweep with respect to x_1. It is easy to see, however, that Eq. (1.5.14) is conditionally stable.

In fact, the coefficient $\varrho(k) = e^{\omega\tau}$ of the harmonic increment is expressed by

$$\varrho(k_1, k_2) = \frac{1 - a_2}{1 + a_1}; \quad a_s = 4 r_s \sin^2 \frac{k_s h_s}{2}; \quad r_s = \frac{a^2 \tau}{h_s^2}; \quad s = 1, 2. \quad (1.5.15)$$

The stability condition is

$$r_2 \leq \frac{1}{2}. \quad (1.5.16)$$

We see that instability could develop as a result of the explicit approximation of the derivative with respect to x_2; while the harmonic solution $A(t)\, e^{ik_1x_1}$ has a steadily decreasing amplitude, the harmonic solution $A(t)\, e^{ik_2x_2}$ will have an increasing amplitude if condition (1.5.16) is not satisfied.

Difficulties encountered in the design of absolutely stable schemes for problems in several dimensions can not be overcome by uniform schemes and simple approximations, when the integration is carried out in a unique manner step by step. A change in the structure of difference scheme is required, and the approximation has to be made more complex.

Let us consider this situation in detail. Up to now we have used the simplest approximations in designing difference schemes. For example, the operator $D = \partial/\partial x$ was approximated by the operator $\alpha \Delta_1/h + (1 - \alpha)\, \Delta_{-1}/h$ and the operator $D^2 = \partial^2/\partial x^2$ by the operator $\Delta_1 \Delta_{-1}/h^2$ etc. As a criterion for the choice of simplicity in the approximation we can require the minimization of the number of network points (for a given accuracy) in the region of determinacy of the difference operator.

Let $\Lambda \sim \Omega$ be the simplest approximation of order $O(h^\alpha)$. Then, putting

$$\bar{\Lambda} = \Lambda + h^\alpha \Phi,$$

where Φ is the arbitrary finite non-singular operator, the approximation $\bar{\Lambda} \sim \Omega$ is also of order $O(h^\alpha)$. In this way we obtain a complete set of arbitrary parameters or functions, related to the arbitrary operator Φ, which could be used in many ways. The use of schemes with a more complicated approximation makes these schemes more flexible and makes it possible to devise schemes which are easily carried out.

Chapter 2

Simple Schemes in Fractional Steps for the Integration of Parabolic Equations

2.1 The scheme of longitudinal-transverse sweep

The conditionally stable scheme (1.5.14) is unsymmetric. The approximation of the second derivative with respect to x is implicit, and with respect to y is explicit. Let us consider a symmetric modification of this scheme in which x and y interchange roles at each step:

$$\begin{aligned} \frac{u^{n+1} - u^n}{\tau} &= \Lambda_1 u^{n+1} + \Lambda_2 u^n, \\ \frac{u^{n+2} - u^{n+1}}{\tau} &= \Lambda_1 u^{n+1} + \Lambda_2 u^{n+2}. \end{aligned} \tag{2.1.1}$$

As in the scheme (1.5.14), at the first step, the operator $L_1 = a^2 \partial^2/\partial x^2$ is approximated implicitly and the operator $L_2 = a^2 \partial^2/\partial y^2$ is approximated explicitly. At the second step, on the other hand, the operator L_1 is approximated explicitly, and the operator L_2 is approximated implicitly. Subsequently, the whole calculation is repeated.

Scheme (2.1.1), which we will call the scheme of longitudinal-transverse sweep or the scheme of alternating directions was proposed at the same time in 1955 by Peaceman, Rachford and Douglas [10, 11]. We shall show that scheme (2.1.1) is absolutely stable and absolutely consistent with the heat conduction equation (1.5.1).

Since the calculation in scheme (2.1.1) is repeated only on going from the n-th step to the $(n + 2)$-th step, we regard the $(n + 1)$-th step as an intermediate step. Therefore, scheme (2.1.1) is considered as a transition from the n-th step to the $(n + 1)$-th step with an auxiliary step $n + 1/2$. With this convention, scheme (2.1.1) has the form

$$\begin{aligned} \frac{u^{n+1/2} - u^n}{\tau} &= \frac{1}{2}\left(\Lambda_1 u^{n+1/2} + \Lambda_2 u^n\right), \\ \frac{u^{n+1} - u^{n+1/2}}{\tau} &= \frac{1}{2}\left(\Lambda_1 u^{n+1/2} + \Lambda_2 u^{n+1}\right). \end{aligned} \tag{2.1.2}$$

We will show that scheme (2.1.2) is equivalent to a certain uniform scheme, unconditionally stable and unconditionally consistent with Eq. (1.5.1). In agreement with [12], Eq. (2.1.2) can be written as follows

$$A_1 u^{n+1/2} - B_1 u^n = 0, \tag{2.1.3a}$$

$$A_2 u^{n+1} - B_2 u^{n+1/2} = 0, \tag{2.1.3b}$$

$$\begin{aligned} A_1 &= E - \frac{1}{2}\tau \Lambda_1, \qquad A_2 = E - \frac{1}{2}\tau\Lambda_2; \\ B_1 &= E + \frac{1}{2}\tau \Lambda_2, \qquad B_2 = E + \frac{1}{2}\tau \Lambda_1. \end{aligned} \tag{2.1.4}$$

Let us multiply Eq. (2.1.3a) by the operator B_2, Eq. (2.1.3b) by A_1, and add, then

$$A_1 A_2 u^{n+1} - B_2 B_1 u^n + (B_2 A_1 - A_1 B_2) u^{n+1/2} = 0.$$

Assuming that the operators Λ_1, Λ_2 are commutative, we obtain the scheme

$$A_1 A_2 u^{n+1} - B_1 B_2 u^n = 0. \tag{2.1.5}$$

Substituting Eq. (2.1.4) into Eq. (2.1.5) we obtain, after some simple transformations, the following uniform scheme, equivalent to scheme (2.1.2)

$$\frac{u^{n+1} - u^n}{\tau} = \frac{\Lambda_1 + \Lambda_2}{2}(u^n + u^{n+1}) - \frac{1}{4}\tau \Lambda_1 \Lambda_2 (u^{n+1} - u^n). \tag{2.1.6}$$

It follows from this that scheme (2.1.6) and the equivalent scheme (2.1.2) approximates the heat conduction equation with the same accuracy as the scheme

$$\frac{u^{n+1} - u^n}{\tau} = \Lambda \frac{u^n + u^{n+1}}{2}, \quad \Lambda = \Lambda_1 + \Lambda_2.$$

Let us prove the unconditional stability of scheme (2.1.6) or, what is equivalent, scheme (2.1.2). Assume that

$$u^n = \eta_n e^{i(k_1 x_1 + k_2 x_2)}; \qquad u^{n+1/2} = \eta_{n+1/2} e^{i(k_1 x_1 + k_2 x_2)}. \tag{2.1.7}$$

Substituting Eq. (2.1.7) into Eq. (2.1.2) we obtain

$$\varrho_1 = \frac{\eta_{n+1/2}}{\eta_n} = \frac{1 - \frac{1}{2} a_2}{1 + \frac{1}{2} a_1}, \tag{2.1.8a}$$

$$\varrho_2 = \frac{\eta_{n+1}}{\eta_{n+1/2}} = \frac{1 - \frac{1}{2} a_1}{1 + \frac{1}{2} a_2}, \tag{2.1.8b}$$

$$\varrho = \frac{1 - \frac{1}{2} a_1}{1 + \frac{1}{2} a_2} \cdot \frac{1 - \frac{1}{2} a_2}{1 + \frac{1}{2} a_1} = \varrho_1 \varrho_2, \tag{2.1.8c}$$

where

$$a_s = 4r_s \sin^2 \frac{k_s h_s}{2}, \quad r_s = \frac{a^2 \tau}{h_s^2}, \quad s = 1, 2. \tag{2.1.9}$$

It follows from this that

$$|\varrho| \leq 1 \tag{2.1.10}$$

for any τ. The stability of the scheme (2.1.2) is proved. It is not difficult to establish that Eq. (2.1.6) gives the same expression for ϱ. In this way, due to the introduction of intermediate fractional steps, we have obtained an absolutely stable scheme. At the same time, instead of one matrix sweep we need two ordinary sweeps and this considerably reduces the extent of the calculations.

Let us analyze formula (2.1.8). Eq. (2.1.8) shows that during the first half step the error in the direction x_1 decreases by $(1 + a_1/2)$ times, and the error in the direction x_2 increases by $(1 - a_2/2)$ times. At the second half step, on the other hand, the error in the direction x_1 increases by $(1 - a_1/2)$ times and that in the direction x_2 decreases by $(1 + a_2/2)$ times. Consequently, no matter how much the error increases in a certain direction at a given half step it inevitably decreases at the next half step so that its modulus does not increase after two half steps. This immediately shows the advantage of the alternating directions scheme over scheme (1.5.14)

$$\frac{u^{n+1} - u^n}{\tau} = \Lambda_1 u^{n+1} + \Lambda_2 u^n$$

and the analogous scheme

$$\frac{u^{n+1} - u^n}{\tau} = \Lambda_1 u^n + \Lambda_2 u^{n+1}.$$

In the first scheme the error in the x_1 direction will always decrease by $(1 + a_1)$ times, but the error in the x_2 direction will always increase by $(1 - a_2)$ times. In the second scheme, on the other hand, the error in the x_1 direction increases by $(1 - a_1)$ times at each step, and it decreases by $(1 + a_2)$ times in the x_2 direction.

Consequently, it is necessary to interchange x_1 and x_2 directions which is done in the alternating directions scheme. In the method of alternating implicit calculation the integration in each direction is carried out alternately by an explicit, and then by an implicit scheme, and the error increase in the explicit scheme is balanced by the error decrease in the implicit scheme[1].

---

[1] The compensation of stability in the fractional steps is analogous to the compensation of binding strength in a plywood panel formed from a series of glued sheets with alternating grain directions. If the sheets are glued with the grain all in the same direction no strength compensation is obtained.

From these considerations it immediately follows that the method of alternating implicit calculation is not applicable to the three-dimensional case. Let us consider a scheme for the three-dimensional heat conduction equation

$$\frac{\partial u}{\partial t} = a^2 \sum_{i=1}^{3} \frac{\partial^2 u}{\partial x_i^2} \tag{2.1.11}$$

which is analogous to the scheme of alternating directions. In this case, the integration in each direction x_1, x_2, and x_3 occurs implicitly once and explicitly twice. This means that the error increase in the explicit scheme is not balanced by its decrease in the implicit scheme. Let us verify this by a precise analysis of the stability of the scheme of alternating directions in the three-dimensional case

$$\left.\begin{aligned}
\frac{u^{n+1/3} - u^n}{\tau} &= \frac{1}{3}\left(\Lambda_1 u^{n+1/3} + \Lambda_2 u^n + \Lambda_3 u^n\right), \\
\frac{u^{n+2/3} - u^{n+1/3}}{\tau} &= \frac{1}{3}\left(\Lambda_1 u^{n+1/3} + \Lambda_2 u^{n+2/3} + \Lambda_3 u^{n+1/3}\right), \\
\frac{u^{n+1} - u^{n+2/3}}{\tau} &= \frac{1}{3}\left(\Lambda_1 u^{n+2/3} + \Lambda_2 u^{n+2/3} + \Lambda_3 u^{n+1}\right).
\end{aligned}\right\} \tag{2.1.12}$$

For the amplification factors we obtain

$$\begin{gathered}
\varrho_1 = \frac{1 - \frac{1}{3}(a_2 + a_3)}{1 + \frac{1}{3} a_1}, \quad \varrho_2 = \frac{1 - \frac{1}{3}(a_1 + a_3)}{1 + \frac{1}{3} a_2}, \quad \varrho_3 = \frac{1 - \frac{1}{3}(a_1 + a_2)}{1 + \frac{1}{3} a_3}; \\
\varrho = \varrho_1 \varrho_2 \varrho_3 = \frac{\left[1 - \frac{1}{3}(a_2 + a_3)\right]\left[1 - \frac{1}{3}(a_1 + a_3)\right]\left[1 - \frac{1}{3}(a_1 + a_2)\right]}{\left(1 + \frac{1}{3} a_1\right)\left(1 + \frac{1}{3} a_2\right)\left(1 + \frac{1}{3} a_3\right)}.
\end{gathered} \tag{2.1.13}$$

It follows immediately that the scheme is not absolutely stable. In fact, for sufficiently large τ/h_i^2, and $i = 1, 2, 3$ the following value is obtained for ϱ

$$\varrho \approx -8. \tag{2.1.14}$$

We note also that the method of alternating directions is not applicable to the solution of the equation

$$\frac{\partial u}{\partial t} = \sum_{i,j=1}^{m} a_{ij} \frac{\partial^2 u}{\partial x_i \, \partial x_j} \tag{2.1.15}$$

even when $m = 2$.

2.2 The scheme of stabilizing corrections

For the solution of the three-dimensional heat conduction equation, Douglas and Rachford [12] proposed the following scheme

$$\left.\begin{aligned} \frac{u^{n+1/3}-u^n}{\tau} &= \Lambda_1 u^{n+1/3} + \Lambda_2 u^n + \Lambda_3 u^n; \\ \frac{u^{n+2/3}-u^{n+1/3}}{\tau} &= \Lambda_2 (u^{n+2/3} - u^n); \\ \frac{u^{n+1}-u^{n+2/3}}{\tau} &= \Lambda_3 (u^{n+1} - u^n). \end{aligned}\right\} \quad (2.2.1)$$

Scheme (2.2.1) can be written in the form

$$A_s u^{n+s/3} - B_s u^{n+(s-1)/3} = C_s u^n, \quad (2.2.2)$$

where

$$\begin{gathered} A_s = E - \tau \Lambda_s; \quad B_s = E; \quad s = 1, 2, 3; \\ C_1 = \tau(\Lambda_2 + \Lambda_3); \quad C_2 = -\tau \Lambda_2; \quad C_3 = -\tau \Lambda_3. \end{gathered} \quad (2.2.2')$$

By successively eliminating $u^{n+1/3}$, $u^{n+2/3}$ from these equations we obtain the equivalent uniform scheme

$$A_1 A_2 A_3 u^{n+1} - B_1 B_2 B_3 u^n = [C_1 + A_1 C_2 + A_1 A_2 C_3] u^n. \quad (2.2.3)$$

After substitution of Eq. (2.2.2′) into Eq. (2.2.3) and expansion with respect to τ, scheme (2.2.3) takes the form

$$\begin{aligned} \frac{u^{n+1}-u^n}{\tau} &= \Lambda u^{n+1} - \tau(\Lambda_1 \Lambda_2 + \Lambda_1 \Lambda_3 + \Lambda_2 \Lambda_3)(u^{n+1} - u^n) + \\ &+ \tau^2 \Lambda_1 \Lambda_2 \Lambda_3 (u^{n+1} - u^n), \end{aligned} \quad (2.2.4)$$

where

$$\Lambda = \Lambda_1 + \Lambda_2 + \Lambda_3.$$

The amplification factor is then given by

$$\varrho = \frac{1 + a_1 a_2 + a_1 a_3 + a_2 a_3 + a_1 a_2 a_3}{(1 + a_1)(1 + a_2)(1 + a_3)}. \quad (2.2.5)$$

Consistency follows from Eq. (2.2.4) and stability from Eq. (2.2.5).

We note that the structure of the scheme is the following: the first fractional step produces absolute consistency with the equation of heat conduction, and all succeeding fractional steps are corrections and serve to improve the stability. For this reason we call this a scheme with stabilizing corrections. Later, J. Douglas [26] proposed a scheme of stabilizing corrections with second order accuracy.

2.3 The splitting scheme for the equation of heat conduction without a mixed derivative (orthogonal system of coordinates)

Analysis of the stability of alternating implicit schemes shows that approximation with an explicit operator reduces the stability of the scheme. This suggests the idea of using only implicit operators at each fractional step. With this in view, one part of the following operator on the right is approximated at each fractional step:

$$L_s = a^2 \frac{\partial^2}{\partial x_s^2}, \tag{2.3.1}$$

the complete approximation being made only at the end of a whole step. A scheme of this kind was first suggested by the author in [13]. We call such schemes splitting schemes. The simplest splitting scheme for the heat conduction equation in three dimensions is

$$\frac{u^{n+1/3} - u^n}{\tau} = \Lambda_1 u^{n+1/3}, \tag{2.3.2a}$$

$$\frac{u^{n+2/3} - u^{n+1/3}}{\tau} = \Lambda_2 u^{n+2/3}, \tag{2.3.2b}$$

$$\frac{u^{n+1} - u^{n+2/3}}{\tau} = \Lambda_3 u^{n+1}. \tag{2.3.2c}$$

Eqs. (2.3.2) can be rewritten in the form

$$A_s u^{n+s/3} - B_s u^{n+(s-1)/3} = 0, \quad A_s = E - \tau \Lambda_s, \quad B_s = E, \quad s = 1, 2, 3. \tag{2.3.3}$$

Eliminating $u^{n+1/3}$, $u^{n+2/3}$ the following equivalent scheme is obtained

$$A_1 A_2 A_3 u^{n+1} - B_1 B_2 B_3 u^n = A_1 A_2 A_3 u^{n+1} - E u^n = 0. \tag{2.3.4}$$

By expanding Eq. (2.3.4) in powers of τ, we obtain

$$\frac{u^{n+1} - u^n}{\tau} = \Lambda u^{n+1} - \tau(\Lambda_1\Lambda_2 + \Lambda_1\Lambda_3 + \Lambda_2\Lambda_3) u^{n+1} + \tau^2 \Lambda_1\Lambda_2\Lambda_3 u^{n+1}. \tag{2.3.5}$$

For ϱ_1, ϱ_2, ϱ_3 we obtain the values

$$\varrho_1 = \frac{1}{1 + a_1}; \quad \varrho_2 = \frac{1}{1 + a_2}; \quad \varrho_3 = \frac{1}{1 + a_3}; \quad \varrho = \frac{1}{(1 + a_1)(1 + a_2)(1 + a_3)}. \tag{2.3.6}$$

Consistency of scheme (2.3.2) follows from Eq. (2.3.5) and stability from Eq. (2.3.6). It is easy to see that scheme (2.3.2) satisfies the extremum condition. This is evident from the fact that each of the two-layer schemes (2.3.2a), (2.3.2b) and (2.3.2c) satisfied the extremum property. Consider, for example, the scheme (2.3.2a), by expressing it first in index form but then eliminating the indices corresponding to x_2 and x_3 for simplicity

$$\frac{u_i^{n+1/3} - u_i^n}{\tau} = a^2 \frac{u_{i-1}^{n+1/3} - 2u_i^{n+1/3} + u_{i+1}^{n+1/3}}{h^2}.$$

Solving then with respect to $u_i^{n+1/3}$, we find

$$u_i^{n+1/3} = \frac{r}{1 + 2r} u_{i-1}^{n+1/3} + \frac{1}{1 + 2r} u_i^n + \frac{r}{1 + 2r} u_{i+1}^{n+1/3}.$$

From this follows the extremum property

$$\min\{u_{i-1}^{n+1/3}, u_i^n, u_{i+1}^{n+1/3}\} \leq u_i^{n+1/3} \leq \max\{u_{i-1}^{n+1/3}, u_i^n, u_{i+1}^{n+1/3}\}.$$

From this condition we deduce, in particular, the convergence in C of the difference solution to the solution of the differential equation (uniform convergence). To improve the accuracy of scheme (2.3.2) the following weighted scheme could be used

$$\left.\begin{aligned} \frac{u^{n+1/3} - u^n}{\tau} &= \Lambda_1[\alpha\, u^{n+1/3} + (1-\alpha)\, u^n]; \\ \frac{u^{n+2/3} - u^{n+1/3}}{\tau} &= \Lambda_2[\alpha\, u^{n+2/3} + (1-\alpha)\, u^{n+1/3}]; \\ \frac{u^{n+1} - u^{n+2/3}}{\tau} &= \Lambda_3[\alpha\, u^{n+1} + (1-\alpha)\, u^{n+2/3}]. \end{aligned}\right\} \tag{2.3.7}$$

The equivalent uniform scheme is

$$\begin{gathered} A_1 A_2 A_3\, u^{n+1} - B_1 B_2 B_3\, u^n = 0, \\ A_s = E - \alpha\tau\Lambda_s, \quad B_s = E + (1-\alpha)\,\tau\Lambda_s, \quad s = 1, 2, 3. \end{gathered} \tag{2.3.8}$$

Expanding Eq. (2.3.8) in powers of τ, we obtain

$$\frac{u^{n+1} - u^n}{\tau} = \Lambda[\alpha\, u^{n+1} + (1-\alpha)\, u^n] + \tau[\Phi_1\, u^{n+1} + \Phi_0\, u^n], \tag{2.3.9}$$

where

$$\begin{aligned} \Phi_1 &= -\alpha^2(\Lambda_1\Lambda_2 + \Lambda_1\Lambda_3 + \Lambda_2\Lambda_3) + \tau\alpha^3\Lambda_1\Lambda_2\Lambda_3; \\ \Phi_0 &= (1-\alpha)^2(\Lambda_1\Lambda_2 + \Lambda_1\Lambda_3 + \Lambda_2\Lambda_3) + \tau(1-\alpha)^3\Lambda_1\Lambda_2\Lambda_3. \end{aligned} \tag{2.3.10}$$

For $\alpha = 1/2$ we get the scheme

$$\begin{aligned} \frac{u^{n+1} - u^n}{\tau} = \Lambda\,\frac{u^{n+1} + u^n}{2} - \frac{\tau^2}{4}(\Lambda_1\Lambda_2 + \Lambda_1\Lambda_3 + \Lambda_2\Lambda_3)\,\frac{u^{n+1} - u^n}{\tau} &+ \\ + \frac{\tau^2}{8}\Lambda_1\Lambda_2\Lambda_3(u^{n+1} + u^n). & \end{aligned} \tag{2.3.11}$$

Scheme (2.3.11) has accuracy of order $O(\tau^2 + h^2)$.

2.4 The splitting scheme for the equation of heat conduction with a mixed derivative (arbitrary system of coordinates)

Consider the equation of parabolic type

$$\frac{\partial u}{\partial t} = L\,u; \quad L = \sum_{i,j=1}^{2} a_{ij}\,\frac{\partial^2}{\partial x_i\,\partial x_j}; \quad a_{ij} = \text{const}, \tag{2.4.1}$$

$$a_{11}\,a_{22} - a_{12}^2 > 0, \quad a_{11} > 0, \quad a_{22} > 0. \tag{2.4.2}$$

In this case the uniform difference scheme

$$\frac{u^{n+1} - u^n}{\tau} = \Lambda[\alpha\, u^{n+1} + (1-\alpha)\, u^n], \quad \Lambda \sim L, \tag{2.4.3}$$

is of nine-point type, and the solution of the system by matrix iteration is very cumbersome. The application of the alternating implicit scheme also does not lead to a simple three-point scheme.

V. A. Suchkov, Yu. Ya. Pogodin and the author [14] suggested the following scheme

$$\frac{u^{n+1/2} - u^n}{\tau} = \Lambda_{11} u^{n+1/2} + \Lambda_{12} u^n;$$
$$\frac{u^{n+1} - u^{n+1/2}}{\tau} = \Lambda_{21} u^{n+1/2} + \Lambda_{22} u^{n+1}, \tag{2.4.4}$$

where

$$\left.\begin{aligned} \Lambda_{11} &= a_{11}\frac{\Delta_1\Delta_{-1}}{h_1^2} \sim L_{11} = a_{11}\frac{\partial^2}{\partial x_1^2};\\ \Lambda_{12} = \Lambda_{21} &= a_{12}\frac{(\Delta_1+\Delta_{-1})(\Delta_2+\Delta_{-2})}{4h_1 h_2} \sim L_{12} = a_{12}\frac{\partial^2}{\partial x_1\,\partial x_2};\\ \Lambda_{22} &= a_{22}\frac{\Delta_2\Delta_{-2}}{h_2^2} \sim L_{22} = a_{22}\frac{\partial^2}{\partial x_2^2}. \end{aligned}\right\} \tag{2.4.5}$$

It is easy to see that scheme (2.4.4) is based on the method of splitting. In fact, at the first half step, "half" of the operator L, namely $L_{11} + L_{12}$, is approximated; L_{11} being approximated in the upper layer $n + 1/2$ and L_{12} in the lower layer. The second "half" of the operator L, namely $L_{21} + L_{22}$ is approximated at the second half-step, $L_{21} = L_{12}$ being approximated in the lower layer $n + 1/2$, and L_{22} in the upper layer $n + 1$.

The equivalent uniform scheme has the form

$$A_{11} A_{22} u^{n+1} - A_{12}^2 u^n = 0; \tag{2.4.6}$$
$$A_{ij} = E + (-1)^{i+j+1}\tau\Lambda_{ij}, \qquad i, j = 1, 2.$$

By expanding in powers of τ, we obtain

$$\frac{u^{n+1} - u^n}{\tau} = (\Lambda_{11} + \Lambda_{22}) u^{n+1} + 2\Lambda_{12} u^n - \tau(\Lambda_{11}\Lambda_{22} u^{n+1} - \Lambda_{12}^2 u^n). \tag{2.4.7}$$

It follows that scheme (2.4.4) is consistent with Eq. (2.4.1). It is easy to prove the stability of the scheme. In fact

$$\varrho_1 = \frac{1 - l_{12}}{1 + l_{11}}; \qquad \varrho_2 = \frac{1 - l_{12}}{1 + l_{22}}; \qquad \varrho = \varrho_1\varrho_2 = \frac{(1 - l_{12})^2}{(1 + l_{11})(1 + l_{22})}, \tag{2.4.8}$$

where

$$l_{ii} = 4\tau\frac{a_{ii}}{h_i^2}\sin^2\frac{k_i h_i}{2};$$
$$l_{12} = 4\tau\frac{a_{12}}{h_1 h_2}\cos\frac{k_1 h_1}{2}\cos\frac{k_2 h_2}{2}\sin\frac{k_1 h_1}{2}\sin\frac{k_2 h_2}{2}. \tag{2.4.9}$$

Using Eq. (2.4.2) we obtain

$$|\varrho| \leq 1. \tag{2.4.10}$$

Stability together with the convergence of the scheme are proved.

For conditions more severe than the elliptic conditions (2.4.2) the method of splitting can also be applied to the three-dimensional heat conduction equation

$$\frac{\partial u}{\partial t} = \sum_{i,j=1}^{3} a_{ij} \frac{\partial^2 u}{\partial x_i \partial x_j}. \tag{2.4.11}$$

In this case we can use the following scheme, proposed by the author [15, 18]

$$\left.\begin{aligned}
\frac{u^{n+1/6} - u^n}{\tau} &= \frac{1}{2}\Lambda_{11} u^{n+1/6} + \Lambda_{12} u^n;\\
\frac{u^{n+2/6} - u^{n+1/6}}{\tau} &= \Lambda_{21} u^{n+1/6} + \frac{1}{2}\Lambda_{22} u^{n+2/6};\\
\frac{u^{n+3/6} - u^{n+2/6}}{\tau} &= \frac{1}{2}\Lambda_{11} u^{n+3/6} + \Lambda_{13} u^{n+2/6};\\
\frac{u^{n+4/6} - u^{n+3/6}}{\tau} &= \Lambda_{31} u^{n+3/6} + \frac{1}{2}\Lambda_{33} u^{n+4/6};\\
\frac{u^{n+5/6} - u^{n+4/6}}{\tau} &= \frac{1}{2}\Lambda_{22} u^{n+5/6} + \Lambda_{23} u^{n+4/6};\\
\frac{u^{n+1} - u^{n+5/6}}{\tau} &= \Lambda_{32} u^{n+5/6} + \frac{1}{2}\Lambda_{33} u^{n+1}.
\end{aligned}\right\} \tag{2.4.12}$$

Scheme (2.4.12) is consistent with expression (2.4.11) and is stable, providing the matrix $\|b_{ij}\|$ is positive definite, where

$$b_{ij} = a_{ij}, \quad i \neq j, \quad b_{ii} = \frac{a_{ii}}{2}. \tag{2.4.13}$$

I. D. Sofronov [65, 66] proposed several integration schemes for Eq. (2.4.1) which are based on the predictor-corrector principle (see Sec. 2.7).

2.5 The scheme of factorization of a difference operator

In the paper of G. A. Baker and T. A. Oliphant [16] the following method of integration of the heat conduction equation (1.5.1) is proposed. Let

$$\Omega u^{n+1} = f^n \tag{2.5.1}$$

be some implicit scheme of integration of Eq. (1.5.1); Ω is a difference operator at the upper step; f^n is the result of the application of difference operators at lower steps. When expressed in suffix notation, scheme (2.5.1) is

$$\sum_{k,l} C_{ij}^{kl} u_{kl}^{n+1} = f_{ij}^n. \tag{2.5.2}$$

It is proved in [16] that if we restrict ourselves to nine-point operators, i.e., operators which satisfy the relation

$$C_{ij}^{kl} = 0 \quad \text{for} \quad |i-k| > 1, \quad |j-l| > 1, \tag{2.5.3}$$

then the operator Ω can be chosen so that it is represented as a product of two three-point operators A and B. This means

$$C_{ij}^{kl} = A_i^k B_j^l, \tag{2.5.4}$$

where

$$A_i^k = 0, \quad |i-k| > 1; \quad B_j^l = 0, \quad |j-l| > 1. \tag{2.5.5}$$

In [16] scheme (2.5.1) is deduced from a three-layer approximation of the heat conduction equation (1.5.1)

$$\frac{1.5u^{n+1} - 2u^n + 0.5u^{n-1}}{\tau} = \Lambda u^{n+1}, \tag{2.5.6}$$

where Λ is some nine-point operator. Then

$$\Omega = 1.5E - \tau\Lambda; \quad f^n = 2u^n - 0.5u^{n-1}. \tag{2.5.7}$$

The operator Λ is chosen so that the approximation

$$\Lambda \sim L = L_1 + L_2 = a^2 \frac{\partial^2}{\partial x^2} + a^2 \frac{\partial^2}{\partial y^2}$$

is of second order accuracy and the operator Ω in expression (2.5.7) is represented as a product of two three-point operators in the sense of the expression (2.5.4). It turns out that with these conditions, Λ is determined uniquely. The solution of the system of Eqs. (2.5.2) is then reduced to two iterations.

In fact, we assume

$$\begin{gathered} A = \{A_i^k\}; \quad B = \{B_i^k\}; \quad v^{n+1} = \{v_{ij}^{n+1}\}; \\ v_{jl}^{n+1} = B_j^k u_{kl}^{n+1} \quad (v^{n+1} = B u^{n+1}). \end{gathered} \tag{2.5.8}$$

Then Eq. (2.5.1) splits into two equations

$$A v^{n+1} = f^n, \quad B u^{n+1} = v^{n+1}, \tag{2.5.9}$$

each of which is then solved by three-point iteration.

In the work of Baker [17] a generalization of the scheme of factorization of the upper operator is given in the case of the equation of heat conduction in several dimensions with constant coefficients. As V. A. Suchkov was kind enough to point out to the author, the scheme of operator factorization suggested by Baker and Oliphant [16] is analogous to the splitting scheme and reduces to it completely when the three-layer approximation (2.5.6) is replaced by the ordinary two-layer approximation.

2.6 The scheme of approximate factorization of operators

In the author's work [18] the method of approximate factorization of operators was described for the heat conduction equation. Let

$$\frac{u^{n+1}-u^n}{\tau} = \Lambda u^{n+1}; \quad \Lambda = \sum_{i=1}^{m} \Lambda_i; \quad \Lambda_i = a^2 \frac{\Delta_i \Delta_{-i}}{h_i^2} \tag{2.6.1}$$

be a simple implicit approximation of the equation

$$\frac{\partial u}{\partial t} = a^2 \sum_{i=1}^{m} \frac{\partial^2 u}{\partial x_i^2}.$$

Eq. (2.6.1) can be written in the form

$$(E - \tau \Lambda) u^{n+1} = E u^n. \tag{2.6.2}$$

Let us factorize the operator $E - \tau \Lambda$ with an accuracy of order τ^2. For this the operator $E - \tau \Lambda$ is replaced by the factorized operator

$$(E - \tau \Lambda_1)(E - \tau \Lambda_2) \ldots (E - \tau \Lambda_m) = E - \tau \Lambda + \tau^2 \Phi,$$

where

$$\Phi = \sum_{i<j} \Lambda_i \Lambda_j - \tau \sum_{i<j<k} \Lambda_i \Lambda_j \Lambda_k + \cdots + (-1)^m \tau^{m-2} \Lambda_1 \ldots \Lambda_m.$$

The difference scheme (2.6.2) is replaced by the factorized scheme

$$\Omega u^{n+1} = (E - \tau \Lambda_1)(E - \tau \Lambda_2) \ldots (E - \tau \Lambda_m) u^{n+1} = E u^n. \tag{2.6.3}$$

Let us introduce the auxiliary quantities $u^{n+1/m}, \ldots, u^{n+(m-1)/m}$ by means of the relations

$$\left.\begin{aligned} (E - \tau \Lambda_1) u^{n+1/m} &= E u^n; \\ (E - \tau \Lambda_2) u^{n+2/m} &= E u^{n+1/m}; \\ \cdots\cdots\cdots & \cdots\cdots\cdots \\ (E - \tau \Lambda_m) u^{n+1} &= E u^{n+(m-1)/m}. \end{aligned}\right\} \tag{2.6.4}$$

The splitting scheme (2.6.4) is equivalent to the scheme of approximate factorization of the upper operator (2.6.3).

The factorization of the upper operator in the scheme of Baker and Oliphant is carried out in a similar way.

Let us consider a three-layer approximation of the two-dimensional heat conduction equation

$$\frac{1.5 u^{n+1} - 2u^n + 0.5 u^{n-1}}{\tau} = \Lambda u^{n+1}, \tag{2.6.5}$$

where

$$\Lambda = \Lambda_1 + \Lambda_2, \quad \Lambda_i \sim L_i = a^2 \frac{\partial^2}{\partial x_i^2}. \tag{2.6.6}$$

The scheme (2.6.5) can be written

$$(1.5 E - \tau \Lambda) u^{n+1} = f^n, \quad f^n = 2u^n - 0.5 u^{n-1}. \tag{2.6.7}$$

Replacing the operator $1.5E - \tau\Lambda$ by a factorized operator, we find

$$\Omega u^{n+1} = 1.5\left(E - \frac{\tau}{1.5}\Lambda_1\right)\left(E - \frac{\tau}{1.5}\Lambda_2\right)u^{n+1} = f^n \qquad (2.6.8)$$

with the nine-point factorized operator Ω, which is identical with the scheme of Baker and Oliphant.

Note that an exact factorization of the upper operator Ω, by means of formulas (2.5.2)–(2.5.4), is impossible in the case of the diffusion equation with variable coefficients, because in this case additional iterations are needed [19] resembling the scheme of N. I. Buleev [20]. At the same time, the method of approximate factorization is equally valid for the equation with variable coefficients.

The method of designing schemes with factorized upper operator was developed for a wide class of equations with variable coefficients by E. G. D'yakonov [21–24]. With the aid of *a priori* estimates he proved the convergence of this method. He also gave an algorithm for the solution of boundary conditions[1]. The method of approximate factorization is discussed in more detail in Sec. 9.3.

2.7 The predictor-corrector scheme

As was shown in Sec. 2.1 and 2.2 the scheme of alternating directions is not applicable in the three-dimensional case because, although it has second order accuracy, it is only conditionally stable. The scheme of stabilizing corrections on the other hand, while being absolutely stable, only has an accuracy of the first order in t.

P. L. I. Brian [25] has proposed an absolutely stable scheme carried out with three-point sweeps, of second order accuracy with respect to t and with respect to space variables. The scheme is derived from the stabilizing corrections scheme and has been called a predictor-corrector process (recalculation).

Let us explain the predictor-corrector method in the simple case of integration of the ordinary differential equation

$$\frac{dx}{dt} = f(x, t). \qquad (2.7.1)$$

The ordinary trapezoidal scheme

$$\frac{x^{n+1} - x^n}{\tau} = \frac{f(x^n, t^n) + f(x^{n+1}, t^{n+1})}{2} \qquad (2.7.2a)$$

or

$$\frac{x^{n+1} - x^n}{\tau} = f\left(\frac{x^{n+1} + x^n}{2}, t^{n+1/2}\right) \qquad (2.7.2b)$$

[1] E. G. D'yakonov used the term "the method of split operators".

requires iteration because of the non-linearity of the right-hand side. The predictor-corrector scheme

$$\frac{x^{n+1/2}-x^n}{\tau/2}=f(x^n,t^n), \tag{2.7.3a}$$

$$\frac{x^{n+1}-x^n}{\tau}=f(x^{n+1/2},t^{n+1/2}) \tag{2.7.3b}$$

is of second-order accuracy and does not require iterations.

In a similar way this process may be applied to partial differential equations. In [25] an integration scheme for Eq. (2.1.11) is proposed which is based on the predictor-corrector principle. This scheme is

$$\frac{u^{n+1/6}-u^n}{\tau/2}=\Lambda_1 u^{n+1/6}+\Lambda_2 u^n+\Lambda_3 u^n, \tag{2.7.4a}$$

$$\frac{u^{n+2/6}-u^{n+1/6}}{\tau/2}=\Lambda_2(u^{n+2/6}-u^n), \tag{2.7.4b}$$

$$\frac{u^{n+3/6}-u^{n+2/6}}{\tau/2}=\Lambda_3(u^{n+3/6}-u^n), \tag{2.7.4c}$$

$$\frac{u^{n+1}-u^n}{\tau}=\Lambda_1 u^{n+1/6}+\Lambda_2 u^{n+2/6}+\Lambda_3 u^{n+3/6}. \tag{2.7.4d}$$

Eqs. (2.7.4a–c) represent the predictor (stabilizing corrections scheme) determining u at $t=(n+1/2)\tau$; (2.7.4d) is the error.

Let us show that scheme (2.7.4) is absolutely stable and of second-order accuracy with respect to t, x_1, x_2 and x_3. Eliminating the fractional steps $u^{n+1/6}$, $u^{n+2/6}$, $u^{n+3/6}$ from Eq. (2.7.4a–d) we have

$$\frac{u^{n+1}-u^n}{\tau}=\Lambda\frac{u^n+u^{n+1}}{2}-\frac{\tau^2}{4}(\Lambda_1\Lambda_2+\Lambda_1\Lambda_3+\Lambda_2\Lambda_3)\frac{u^{n+1}-u^n}{\tau}+$$

$$+\frac{\tau^3}{8}\Lambda_1\Lambda_2\Lambda_3\frac{u^{n+1}-u^n}{\tau}. \tag{2.7.5}$$

The absolute stability and second-order accuracy of scheme (2.7.4) follows from this expression.

J. Douglas [26] has proposed a scheme in fractional steps which is also absolutely stable and of second-order accuracy:

$$\left.\begin{aligned}
\frac{1}{2}\Lambda_1(u^{n+1/3}+u^n)+\Lambda_2 u^n+\Lambda_3 u^n&=\frac{u^{n+1/3}-u^n}{\tau};\\
\frac{1}{2}\Lambda_1(u^{n+1/3}+u^n)+\frac{1}{2}\Lambda_2(u^{n+2/3}+u^n)+\Lambda_3 u^n&=\frac{u^{n+2/3}-u^n}{\tau};\\
\frac{1}{2}\Lambda_1(u^{n+1/3}+u^n)+\frac{1}{2}\Lambda_2(u^{n+2/3}+u^n)+\frac{1}{2}\Lambda_3(u^{n+1}+u^n)&\\
&=\frac{u^{n+1}-u^n}{\tau}.
\end{aligned}\right\} \tag{2.7.6}$$

Scheme (2.7.6) can be expressed in the form

$$\left.\begin{aligned}\frac{u^{n+1/3}-u^n}{\tau} &= \frac{1}{2}\Lambda_1(u^{n+1/3}+u^n)+\Lambda_2 u^n+\Lambda_3 u^n;\\ \frac{u^{n+2/3}-u^{n+1/3}}{\tau} &= \frac{1}{2}\Lambda_2(u^{n+2/3}-u^n);\\ \frac{u^{n+1}-u^{n+2/3}}{\tau} &= \frac{1}{2}\Lambda_3(u^{n+1}-u^n),\end{aligned}\right\} \tag{2.7.7}$$

whence it follows that this scheme is of the stabilizing corrections type. The scheme after eliminating the intermediate steps reduces to scheme (2.7.5); i.e., schemes (2.7.4) and (2.7.6) are equivalent.

Let us point out one more predictor-corrector scheme

$$\frac{u^{n+1/6}-u^n}{\tau/2} = \Lambda_1 u^{n+1/6}, \tag{2.7.8a}$$

$$\frac{u^{n+2/6}-u^{n+1/6}}{\tau/2} = \Lambda_2 u^{n+2/6}, \tag{2.7.8b}$$

$$\frac{u^{n+1/2}-u^{n+2/6}}{\tau/2} = \Lambda_3 u^{n+1/2}, \tag{2.7.8c}$$

$$\frac{u^{n+1}-u^n}{\tau} = \Lambda u^{n+1/2}. \tag{2.7.8d}$$

Formulas (2.7.8a–c) represent the predictor, based on a splitting scheme, and formula (2.7.8d) is the corrector. This scheme for whole steps is

$$\frac{u^{n+1}-u^n}{\tau} = \Lambda\frac{u^n+u^{n+1}}{2}-\left(\frac{\tau}{2}\right)^2(\Lambda_1\Lambda_2+\Lambda_1\Lambda_3+\Lambda_2\Lambda_3)\frac{u^{n+1}-u^n}{\tau}+ \\ +\left(\frac{\tau}{2}\right)^3\Lambda_1\Lambda_2\Lambda_3\frac{u^{n+1}-u^n}{\tau}. \tag{2.7.9}$$

The scheme for whole steps is identical with (2.7.5), i.e., scheme (2.7.8) is equivalent to the preceding schemes.

Schemes which are based on predictor-corrector principle are called schemes of approximation corrections (a.c.).

2.8 Some remarks regarding schemes with fractional steps

The following remarks should be made in regard to the above schemes.

(i) Schemes with fractional steps can be applied to one-dimensional problems. Consider, for example, the locally-implicit scheme of V. K. Saul'ev [27] for solving the one-dimensional heat conduction equation

$$\begin{aligned}\frac{u^{n+1/2}-u^n}{\tau} &= \frac{1}{2}\Lambda u^{n+1/2},\\ \frac{u^{n+1}-u^{n+1/2}}{\tau} &= \frac{1}{2}\Lambda u^{n+1/2}, \quad \Lambda = a^2\frac{\Delta_1\Delta_{-1}}{h^2},\end{aligned} \tag{2.8.1}$$

which uses the alternating directions scheme for one of the directions, for example, x_1. Let us show that scheme (2.8.1) is equivalent to a weighted two-layer scheme for $\alpha = 1/2$ (the Crank-Nicholson scheme). By expressing (2.8.1) in the form

$$\left(E - \frac{\tau}{2}\Lambda\right)u^{n+1/2} = u^n, \qquad u^{n+1} = \left(E + \frac{\tau}{2}\Lambda\right)u^{n+1/2}, \tag{2.8.2}$$

and eliminating $u^{n+1/2}$, we obtain

$$\left(E - \frac{\tau}{2}\Lambda\right)u^{n+1} = \left(E + \frac{\tau}{2}\Lambda\right)u^n. \tag{2.8.3}$$

It follows that

$$\frac{u^{n+1} - u^n}{\tau} = \Lambda\frac{u^n + u^{n+1}}{2}. \tag{2.8.4}$$

This is the required proof.

(ii) Let us write scheme (2.8.1) with integer indices,

$$\frac{u^n - u^{n-1}}{\tau} = \Lambda u^n, \tag{2.8.5 a}$$

$$\frac{u^{n+1} - u^n}{\tau} = \Lambda u^n. \tag{2.8.5 b}$$

Addition of Eqs. (2.8.5) results in

$$\frac{u^{n+1} - u^{n-1}}{\tau} = 2\Lambda u^n. \tag{2.8.6}$$

The scheme (2.8.6) superficially resemble the "cross" scheme; however, formulas (2.8.5) indicate that scheme (2.8.6) is non-uniform and it can not be considered as a uniform "cross" scheme. To derive a uniform scheme, u^n should be eliminated from Eqs. (2.8.5 a) and (2.8.5 b), and as with expression (2.8.4), we obtain

$$\frac{u^{n+1} - u^{n-1}}{\tau} = \Lambda(u^{n-1} + u^{n+1}). \tag{2.8.7}$$

Thus the same formula (2.8.6) can be used in several different ways depending on how we define u^n.

Let us point out another similar example. The stabilizing corrections scheme

$$\begin{aligned} \frac{u^{n+1/2} - u^n}{\tau} &= \Lambda_1 u^{n+1/2} + \Lambda_2 u^n; \\ \frac{u^{n-1} - u^{n+1/2}}{\tau} &= \Lambda_2(u^{n+1} - u^n) \end{aligned} \tag{2.8.8}$$

can be written, by adding Eqs. (2.8.8),

$$\frac{u^{n+1} - u^n}{\tau} = \Lambda_1 u^{n+1/2} + \Lambda_2 u^{n+1}. \tag{2.8.9}$$

The majorant splitting scheme

$$\frac{u^{n+1/2} - u^n}{\tau} = \Lambda_1 u^{n+1/2};$$
$$\frac{u^{n+1} - u^{n+1/2}}{\tau} = \Lambda_2 u^{n+1} \qquad (2.8.10)$$

also produces Eq. (2.8.9) after addition. However, schemes (2.8.8) and (2.8.10) differ and Eq. (2.8.9) has a different meaning in each scheme because the value $u^{n+1/2}$ is not defined in the same way.

The corresponding schemes in whole steps, which are obtained by eliminating $u^{n+1/2}$, are not identical.

(iii) Identical schemes in whole steps can yield different schemes in fractional steps. For example, the scheme in whole steps

$$\left(E - \frac{1}{2}\tau\Lambda_1\right)\left(E - \frac{1}{2}\tau\Lambda_2\right)u^{n+1} = \left(E + \frac{1}{2}\tau\Lambda_1\right)\left(E + \frac{1}{2}\tau\Lambda_2\right)u^n \qquad (2.8.11)$$

could be used with the following schemes in fractional steps:

(a) The alternating directions (a.d.) scheme

$$\frac{u^{n+1/2} - u^n}{\tau} = \frac{1}{2}\left(\Lambda_1 u^{n+1/2} + \Lambda_2 u^n\right);$$
$$\frac{u^{n+1} - u^{n+1/2}}{\tau} = \frac{1}{2}\left(\Lambda_1 u^{n+1/2} + \Lambda_2 u^{n+1}\right). \qquad (2.8.12)$$

(b) The splitting scheme

$$\frac{u^{n+1/2} - u^n}{\tau} = \Lambda_1(\alpha u^{n+1/2} + \beta u^n);$$
$$\frac{u^{n+1} - u^{n+1/2}}{\tau} = \Lambda_2(\alpha u^{n+1} + \beta u^{n+1/2}), \quad \beta = 1 - \alpha, \qquad (2.8.13)$$

for $\alpha = 1/2$.

(c) The scheme of approximate factorization of an operator

$$\left(E - \frac{1}{2}\tau\Lambda_1\right)u^{n+1/2} = f^n = \left(E + \frac{1}{2}\tau\Lambda_1\right)\left(E + \frac{1}{2}\tau\Lambda_2\right)u^n;$$
$$\left(E - \frac{1}{2}\tau\Lambda_2\right)u^{n+1} = u^{n+1/2}. \qquad (2.8.14)$$

Consequently, schemes (2.8.12) – (2.8.14) could be considered as interpretations of one and the same scheme (2.8.11). When boundary conditions are not considered, the schemes (2.8.12) – (2.8.14) are equivalent.

In the case of an equation with variable coefficients the corresponding schemes in whole steps do not coincide and schemes (2.8.12) – (2.8.14) become non-equivalent. In addition one and the same scheme with fractional steps could have different interpretations of the boundary conditions.

(iv) The weighted method of splitting for $m \geq 3$, $\alpha = 1/2$ is extremely simple while at the same time preserving strong stability and accuracy of second order. Nevertheless, the application of the predictor-corrector scheme is most expedient when it is necessary to integrate the heat conduction equation in a time interval including the asymptotic regime ($t \to \infty$). In this case the predictor-corrector schemes as opposed to the splitting scheme permit the step size to be increased because they satisfy the condition of total consistency (see Ch. 4).

2.9 Boundary conditions in the method of fractional steps for the heat conduction equation

Up to now we have considered the Cauchy problem in the strip $|x| < \infty$, $0 \le t \le T$ assuming that the initial data are represented by Fourier series or Fourier integrals. For this there were no boundary points of the network, and difference equations were identical at all network points. In practice it is necessary to solve the mixed Cauchy problem when the difference equations at the boundary are different from those at the interior points. As a result the order of magnitude of the remainder terms of the scheme could be different on the boundary than at interior points. Thus, the boundary conditions influence the accuracy of the scheme.

The first analysis of boundary conditions in schemes with fractional steps, for the heat conduction equation, was carried out by E. G. D'yakonov [22][1]. In line with his work let us consider the splitting schemes with three formulations of boundary conditions. For the equation

$$\frac{\partial u}{\partial t} = a^2 \left(\frac{\partial^2 u}{\partial x_1^2} + \frac{\partial^2 u}{\partial x_2^2} \right) \tag{2.9.1}$$

the mixed Cauchy problem is

$$\begin{aligned} u(x_1, x_2, 0) &= u_0(x_1, x_2), \qquad (x_1, x_2) \in G; \\ u(x_1, x_2, t) &= f(x_1, x_2, t), \qquad (x_1, x_2, t) \in \Gamma, \end{aligned} \tag{2.9.1 a}$$

in the cylindrical region $\Pi = G \times \Gamma$ where G is the square; $0 < x_1, x_2 < 1$; γ is its boundary; $H = [0, t_0]$, $\Gamma = \gamma \times H$ (see Fig. 1).

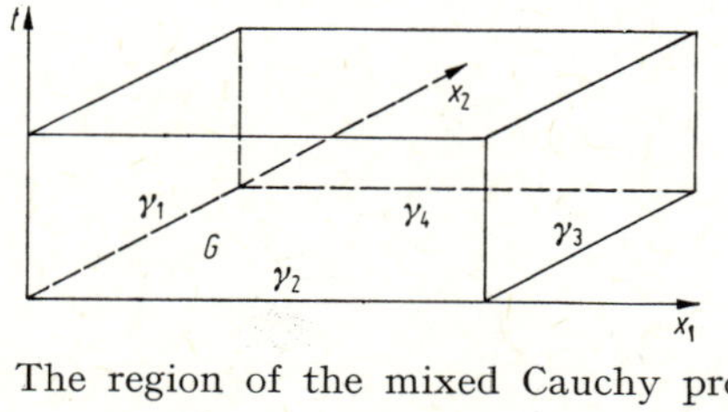

Fig. 1. The region of the mixed Cauchy problem

[1] Unfortunately, this work contains an arithmetical mistake and as a consequence gives an improper evaluation of the method of splitting. Nevertheless, the method of investigating boundary conditions proposed by E. G. D'yakonov, in principle, is correct.

For problem (2.9.1) we shall use the splitting scheme

$$\frac{u^{n+1/2} - u^n}{\tau} = \Lambda_1(\alpha\, u^{n+1/2} + \beta\, u^n), \tag{2.9.2a}$$

$$\frac{u^{n+1} - u^{n+1/2}}{\tau} = \Lambda_2(\alpha\, u^{n+1} + \beta\, u^{n+1/2}), \tag{2.9.2b}$$

and the scheme of approximate operator factorization

$$(E - \alpha\tau\Lambda_1)(E - \alpha\tau\Lambda_2)\, u^{n+1} = (E + \beta\tau\Lambda_1)(E + \beta\tau\Lambda_2)\, u^n \tag{2.9.3}$$

in the following form[1]

$$(E - \alpha\tau\Lambda_1)\, u^{n+1/2} = (E + \beta\tau\Lambda_1)(E + \beta\tau\Lambda_2)\, u^n, \tag{2.9.4a}$$

$$(E - \alpha\tau\Lambda_2)\, u^{n+1} = u^{n+1/2}. \tag{2.9.4b}$$

Consider the following interpretation of the boundary conditions for scheme (2.9.2)

$$\begin{aligned} u^{n+1/2}(x_1, x_2) &= f[x_1, x_2, (n + 1/2)\tau], \quad (x_1, x_2) \in \gamma; \\ u^{n+1}(x_1, x_2) &= f[x_1, x_2, (n + 1)\tau], \quad (x_1, x_2) \in \gamma. \end{aligned} \tag{A}$$

In the case of (A) the sweep is not performed along the boundary, for the values of u at the boundary are always equal to the boundary values at the corresponding instant. It follows from this that at the boundary relations (2.9.2) become

$$\begin{aligned} \frac{u^{n+1/2} - u^n}{\tau} &= \Lambda_1(\alpha\, u^{n+1/2} + \beta\, u^n) + F_1^n; \\ \frac{u^{n+1} - u^{n+1/2}}{\tau} &= \Lambda_2(\alpha\, u^{n+1} + \beta\, u^{n+1/2}) + F_2^n, \end{aligned} \tag{2.9.5}$$

where

$$\begin{aligned} F_1^n &= \frac{f^{n+1/2} - f^n}{\tau} - \Lambda_1(\alpha f^{n+1/2} + \beta f^n), \quad (x_1, x_2) \in \gamma_2, \gamma_4; \\ F_2^n &= \frac{f^{n+1} - f^{n+1/2}}{\tau} - \Lambda_2(\alpha f^{n+1} + \beta f^{n+1/2}), \quad (x_1, x_2) \in \gamma_1, \gamma_3. \end{aligned} \tag{2.9.6}$$

The difference scheme (2.9.2) may therefore be written

$$A_1\, u^{n+1/2} - B_1\, u^n = g_1^n; \tag{2.9.7a}$$

$$A_2\, u^{n+1} - B_2\, u^{n+1/2} = g_2^n, \tag{2.9.7b}$$

where

$$\left.\begin{aligned} A_s &= E - \alpha\tau\Lambda_s, & B_s &= E + \beta\tau\Lambda_s, & s &= 1, 2; \\ g_1^n &= 0, & g_2^n &= 0, & (x_1, x_2) &\in G, \\ g_1^n &= \tau F_1^n, & g_2^n &= \tau F_2^n, & (x_1, x_2) &\in \gamma. \end{aligned}\right\} \tag{2.9.8}$$

The equivalent scheme in whole steps is

$$A_1 A_2\, u^{n+1} - B_1 B_2\, u^n = R_n, \quad R_n = B_2\, g_1^n + A_1\, g_2^n. \tag{2.9.9}$$

[1] Note that schemes (2.9.2), (2.9.3) and (2.9.4) are equivalent for a purely periodic Cauchy problem, but are not equivalent in the case of the boundary problem (2.9.1a). In the case of variable coefficients, scheme (2.9.3) is not in general equivalent to Eq. (2.9.2).

It is clear that $R_n = 0$ everywhere inside G, except at network points located on the line ω, which is separated from γ by one interval (see Fig. 2), where

$$R_n = B_2 g_1^n, \quad (x_1, x_2) \in \omega_2, \omega_4; \qquad R_n = A_1 g_2^n, \quad (x_1, x_2) \in \omega_1, \omega_3. \tag{2.9.10}$$

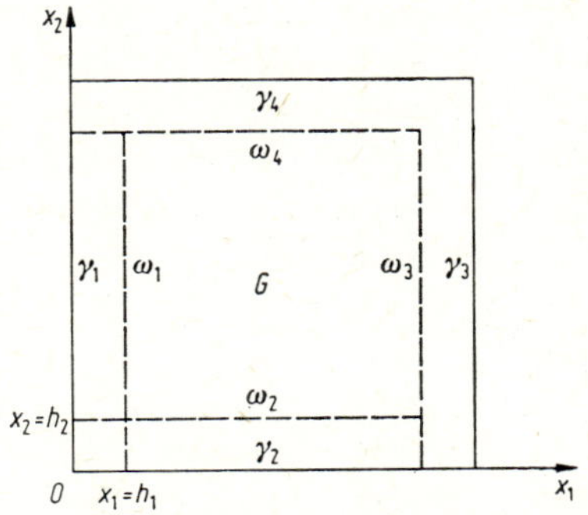

Fig. 2. The boundary γ and the set ω
$(N_1 + 1)\, h_1 = 1, \quad (N_2 + 1)\, h_2 = 1$

Simple calculation gives

$$\begin{aligned} R_n &= B_2 g_1^n = \beta \frac{\tau^2}{h_2^2} F_1^n, && (x_1, x_2) \in \omega_2, \omega_4; \\ R_n &= A_1 g_2^n = -\alpha \frac{\tau^2}{h_1^2} F_2^n, && (x_1, x_2) \in \omega_1, \omega_3. \end{aligned} \tag{2.9.11}$$

It follows from this that when we proceed to the limit according to this law

$$\frac{\tau}{h_i^2} = \text{const}$$

the approximation error at points on ω is $O(1)$, i.e., it is finite. Since this error is equal to $O(\tau)$ at all other interior points of G, then, when the error is evaluated in L_2, we have $\|u\|_{L_2} = O(\tau^{1/4})$.

We see that a large approximation error is connected with the violation of relations (2.9.2) on the boundary. The formulation of (A) was considered in [22].

Consider now another formulation of the boundary conditions in the splitting scheme:

$$\left.\begin{aligned} &\text{(a)} \quad u^{n+1/2}(x_1, x_2) = f^{n+1/2}(x_1, x_2), \quad (x_1, x_2) \in \gamma_1, \gamma_3; \\ &\text{(b)} \quad A_1 u^{n+1/2}(x_1, x_2) - B_1 u^n(x_1, x_2) = A_1 u^{n+1/2} - B_1 f^n = 0, \\ &\qquad\qquad (x_1, x_2) \in \gamma_2, \gamma_4; \\ &\text{(c)} \quad u^{n+1}(x_1, x_2) = f^{n+1}(x_1, x_2), \quad (x_1, x_2) \in \gamma_2, \gamma_4, \\ &\text{(d)} \quad A_2 u^{n+1}(x_1, x_2) - B_2 u^{n+1/2}(x_1, x_2) = A_2 u^{n+1} - B_2 f^{n+1/2} = 0, \\ &\qquad\qquad (x_1, x_2) \in \gamma_1, \gamma_3. \end{aligned}\right\} \text{(B)}$$

In the case of (B), at the first fractional step the boundary conditions are satisfied exactly on γ_1, γ_3; on γ_2, γ_4 they are replaced by conditions (b). At the second fractional step, on the other hand, the boundary conditions are satisfied exactly on γ_2, γ_4, and are replaced by conditions (d) on γ_1, γ_3. The fact that conditions (b) and (d) are satisfied means that sweep is also carried out along the boundary. Using this approach we satisfy the boundary conditions on the boundary at one fractional step, but Eq. (2.9.2) is satisfied everywhere in $\bar{G}$. Consequently, the elimination of fractional steps leads to a scheme in whole steps of type (2.9.3) everywhere in $\bar{G}$; i.e., the order of the approximation is $O(\tau) + O(h^2)$ for $\alpha \neq 1/2$ and $O(\tau^2) + O(h^2)$ for $\alpha = 1/2$. Let us show that in the case of sweep along the boundary we commit an error in the boundary condition of order $O(\tau)$. Let us demonstrate this, for example, for γ_2, γ_4.

At the n-th step we have $u^n = f^n$ on γ_2, γ_4; $u^{n+1/2}$ is determined from the condition (2.9.2a) which can be expressed in the form

$$A_1 u^{n+1/2} = B_1 u^n = B_1 f^n. \tag{2.9.12}$$

The recurrence relation (2.9.12) is closed with the boundary conditions

$$u^{n+1/2} = f^{n+1/2} \tag{2.9.13}$$

at $x_1 = 0, 1$.

Subtracting the identity

$$A_1 f^{n+1/2} = A_1 f^{n+1/2}$$

from relation (2.9.12) and denoting the difference $u^{n+1/2} - f^{n+1/2}$ by $v^{n+1/2}$ we have

$$A_1 v^{n+1/2} = B_1 f^n - A_1 f^{n+1/2} = F^n \tag{2.9.14}$$

with boundary conditions

$$v^{n+1/2} = 0 \tag{2.9.15}$$

at $x_1 = 0, 1$. From (2.9.14) we have

$$v^{n+1/2} = A_1^{-1} F^n.$$

But

$$\|F^n\| = O(\tau), \quad \|A_1^{-1}\| < 1,$$

consequently

$$\|v^{n+1/2}\| = O(\tau). \tag{2.9.16}$$

Thus, when sweeping along x_1, we also perform a sweep along γ_2, γ_4, and depart from the true boundary conditions by $O(\tau)$, while preserving them on γ_1, γ_3. When sweeping along x_2, on the other hand, the boundary conditions on γ_2, γ_4 are preserved while a sweep is performed along γ_1, γ_3 with a departure from actual boundary conditions of $O(\tau)$. The final accuracy is $O(\tau) + O(h^2)$.

Consider, finally, a third method for satisfying the boundary conditions, which corresponds to the interpretation (2.9.4) of scheme (2.9.3). For the sweep of $u^{n+1/2}$ along x_1 it is necessary to know $u^{n+1/2}$ on γ_1, γ_3. This value of $u^{n+1/2}$ is determined from the relation (2.9.4b) which is transformed into

$$u^{n+1/2} = (E - \beta\,\tau\,\Lambda_2)\,f^{n+1} \tag{2.9.17}$$

on γ_1, γ_3.

The calculation is accomplished in the following order: $u^{n+1/2}$ is determined on γ_1, γ_3, from relation (2.9.17), after which Eq. (2.9.4a) is solved by sweep which gives $u^{n+1/2}$ in G. Then Eq. (2.9.4b) is solved by sweep with the assumption that

$$u^{n+1} = f^{n+1} \tag{2.9.18}$$

on γ_2, γ_4.

Consequently, Eq. (2.9.3) is satisfied everywhere in G at whole steps, and the relation

$$u^n = f^n, \qquad u^{n+1} = f^{n+1} \tag{C}$$

is satisfied on γ. The scheme (C) for $\alpha = 1/2$ gives an approximation of order $O(\tau^2) + O(h^2)$.

Compare now the calculation of the boundaries by methods (A), (B) and (C) for the case of a rectangle. Method (A) evidently is of lowest accuracy. Method (C) is superior to method (B) with respect to accuracy. However, method (A) has the advantage of generality in the case of an arbitrary region.

In fact, consider the region G which is confined within an arbitrary contour γ (see Fig. 3). Method (C) is unacceptable in this case. Method (A)

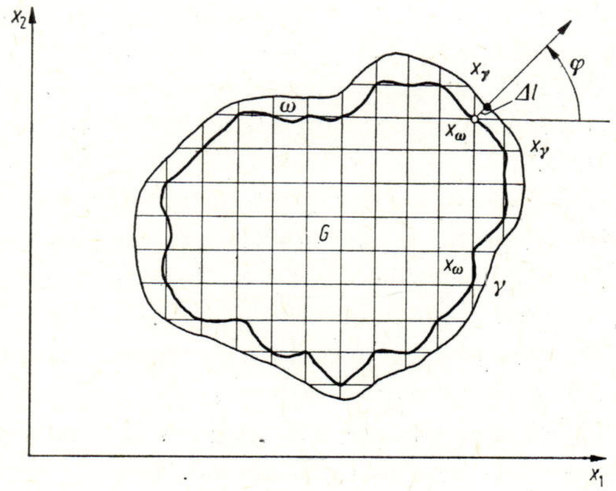

Fig. 3.
The method (A) for the arbitrary region G

is applicable here for the following reasons. For the solution of Eq. (2.9.2a) the sweep is performed along horizontal network lines with boundary conditions at ends of the horizontal sections, while Eq. (2.9.2b) is solved by sweep along vertical network lines with boundary conditions at ends of the vertical sections. Moreover, the ends of the horizontal and vertical sections of the network do not have to coincide; i.e., the boundary does not necessarily contain only network points (non-concordant network). Method (C) is inapplicable even in the case of a region consisting of finite number of rectangles. For example, for the region shown in Fig. 4 the value $u^{n+1/2}$ at angular points A, B, C, D, E, F cannot be

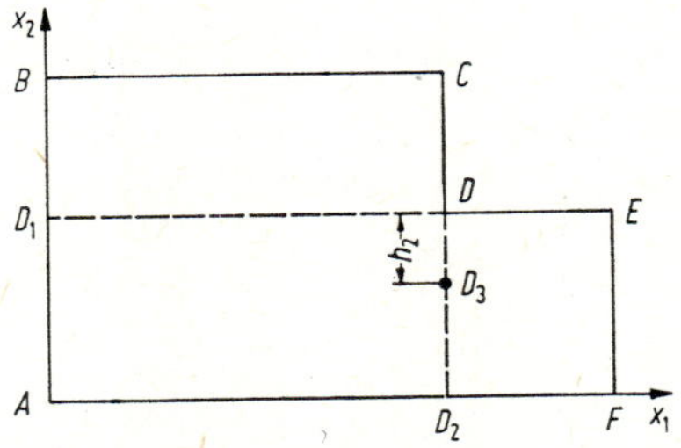

Fig. 4. The application of method (C) in the region G which resembles a rectangle

determined. Consequently, the sweep along DD_1 is impossible. In this case, Method (C) is carried out as follows. With the aid of formulas (2.9.17), $u^{n+1/2}$ is determined on AB, CD, EF.

After this the sweep along x_1 is carried out along all interior horizontal network lines, with the exception of the line D_1D, and $u^{n+1/2}$ is determined everywhere in the region G, with the exception of D_1D. Then the sweep along x_2 in $DEFD_2$ is performed and u^{n+1} is determined everywhere in $DEFD_2$, in particular at the point D_3. With the use of (2.9.17) $u^{n+1/2}$ is determined at point D and the sweep of $u^{n+1/2}$ is performed along D_1D.

After this it becomes possible to determine u^{n+1} in $ABCD_2$ with the use of sweep along x_2. It is evident that with the increase of the number of angular points the algorithm becomes very cumbersome. At this point, method (A) should be applied.

N. N. Anuchina suggested in a private communication a more general method of treating boundary conditions for methods of fractional steps (the method of undetermined functions). Let us demonstrate this method in the case of the splitting scheme. Denote the values of u^n, $u^{n+1/2}$ at the boundary γ by ψ^n and φ^n, respectively. It is evident that the values ψ^n, φ^n do not necessarily coincide with f^n and $f^{n+1/2}$, respectively.

Relations (2.9.2) on the boundary γ take the form of Eq. (2.9.5), where

$$\left.\begin{aligned} F_1^n &= \frac{\varphi^n - \psi^n}{\tau} - \Lambda_1(\alpha\,\varphi^n + \beta\,\psi^n) = \frac{1}{\tau}(A_1\,\varphi^n - B_1\,\psi^n), \\ &\qquad (x_1, x_2) \in \gamma_2, \gamma_4; \\ F_2^n &= \frac{\psi^{n+1} - \varphi^n}{\tau} - \Lambda_2(\alpha\,\psi^{n+1} + \beta\,\varphi^n) = \frac{1}{\tau}(A_2\,\psi^{n+1} - B_2\,\varphi^n), \\ &\qquad (x_1, x_2) \in \gamma_1, \gamma_3. \end{aligned}\right\} \quad (2.9.19)$$

The equivalent scheme in whole steps is

$$A_1\,A_2\,u^{n+1} - B_1\,B_2\,u^n = R_n, \quad (2.9.20)$$

$R_n = 0$ everywhere in G, with the exception of ω, where

$$\left.\begin{aligned} R_n &= B_2(A_1\,\varphi^n - B_1\,\psi^n), \qquad (x_1, x_2) \in \omega_2, \omega_4; \\ R_n &= A_1(A_2\,\psi^{n+1} - B_2\,\varphi^n), \qquad (x_1, x_2) \in \omega_1, \omega_3. \end{aligned}\right\} \quad (2.9.21)$$

The problem consists in choosing the functions ψ^n, φ^n so that the error R_n is minimized. It has already been shown that the trivial choice $\psi^n = f^n$, $\varphi^n = f^{n+1/2}$ leads to method (A), which is of low accuracy.

In order that $R_n = 0$ everywhere in G, including ω, it is necessary and sufficient that

$$A_2\,\psi^{n-1} - B_2\,\varphi^n = 0, \qquad (x_1, x_2) \in \gamma_2, \gamma_4; \quad (2.9.22\text{a})$$

$$A_1\,\varphi^n - B_1\,\psi^n = 0, \qquad (x_1, x_2) \in \gamma_1, \gamma_3. \quad (2.9.22\text{b})$$

If we assume that

$$\psi^n = f^n, \qquad \psi^{n+1} = f^{n+1}, \qquad (x_1, x_2) \in \gamma, \quad (2.9.23)$$

then Eq. (2.9.22) enables us to determine φ^n. However, Eq. (2.9.22b) is badly posed and cannot be solved by sweep.

While retaining assumption (2.9.23) let us choose φ^n so that $R_n = O(\tau^2)$. For this, it is sufficient, for example, to put

$$\left.\begin{aligned} \varphi^n &= (E + k_1\,\tau\,\Lambda_1)\,\psi^n, \qquad (x_1, x_2) \in \omega_2, \omega_4; \\ \varphi^n &= (E + k_2\,\tau\,\Lambda_2)\,\psi^{n+1}, \qquad (x_1, x_2) \in \omega_1, \omega_3. \end{aligned}\right\} \quad (2.9.24)$$

Let us choose the constants k_1, k_2 so that

$$\left.\begin{aligned} R_n &= B_2(A_1\,\varphi^n - B_1\,\psi^n) = O(\tau^2), \qquad (x_1, x_2) \in \omega_2, \omega_4; \\ R_n &= A_1(A_2\,\psi^{n-1} - B_2\,\varphi^n) = O(\tau^2), \qquad (x_1, x_2) \in \omega_1, \omega_3. \end{aligned}\right\} \quad (2.9.25)$$

For $k_1 = 1$, $k_2 = -1$ we have

$$\left.\begin{aligned} R_n &= -\frac{\alpha\,\beta\,\tau}{h_2^2}\,\tau^2\,\Lambda_1^2 f^n, \qquad (x_1, x_2) \in \omega_2, \omega_4; \\ R_n &= -\frac{\alpha\,\beta\,\tau}{h_1^2}\,\tau^2\,\Lambda_2^2 f^{n+1}, \qquad (x_1, x_2) \in \omega_1, \omega_3. \end{aligned}\right\} \quad (2.9.26)$$

Using the law of approach to the limit $\tau/h_1^2 = \text{const}$, $\tau/h_2^2 = \text{const}$, together with a sufficiently smooth function f we achieve the required accuracy. Note, however, that the consistency (2.9.26) is not absolute. In addition, if f^n is a discontinuous function, the order of consistency of Eq. (2.9.26) is reduced.

Assume now, method (B), that

$$\begin{aligned} \varphi^n &= f^{n+1/2}, & (x_1, x_2) &\in \gamma_1, \gamma_3; \\ A_1 \varphi^n - B_1 f^n &= 0, & (x_1, x_2) &\in \gamma_2, \gamma_4; \end{aligned} \tag{2.9.27}$$

$$\begin{aligned} \psi^{n+1} &= f^{n+1}, & (x_1, x_2) &\in \gamma_2, \gamma_4; \\ A_2 \psi^{n+1} - B_1 f^{n+1/2} &= 0, & (x_1, x_2) &\in \gamma_1, \gamma_3. \end{aligned} \tag{2.9.28}$$

As in the case of Eq. (2.9.22), $R_n = 0$, but as opposed to Eq. (2.9.22) ψ^{n+1} rather than φ^n is determined from Eq. (2.9.27).

Formulas (2.9.27) and (2.9.28) produce accuracy of the order of $O(\tau) + O(h^2)$, regardless of the law of approach to the limit and the smoothness of f^n.

Other schemes of fractional steps are derived in an analogous way. The method of undetermined functions can be applied to the cases of curvilinear and concave boundaries, as well as to equations of more complex structure.

The method of undetermined functions can be formulated in a more general way. Consider for example the splitting scheme. Assume that

$$\begin{aligned} \frac{u^{n+1/2} - u^n}{\tau} &= \Lambda_1(\alpha\, u^{n+1/2} + \beta\, u^n) + q_1^n; \\ \frac{u^{n+1} - u^{n+1/2}}{\tau} &= \Lambda_2(\alpha\, u^{n+1} + \beta\, u^{n+1/2}) + q_2^n \end{aligned} \tag{2.9.29}$$

has temporarily undetermined functions q_1 and q_2. By eliminating $u^{n+1/2}$ from (2.9.29) we obtain

$$(E - \alpha\tau\Lambda_1)(E - \alpha\tau\Lambda_2)u^{n+1} = (E + \beta\tau\Lambda_1)(E + \beta\tau\Lambda_2)u^n + R, \tag{2.9.30}$$

$$R = \tau[(E + \beta\,\tau\,\Lambda_2)\, q_1 + (E - \alpha\,\tau\,\Lambda_1)\, q_2]. \tag{2.9.31}$$

Let us require that

$$\begin{aligned} R &= 0, & (x_1, x_2) &\in G; \\ q_1 &= 0, & (x_1, x_2) &\in G. \end{aligned} \tag{2.9.32}$$

Let us show that conditions (2.9.32) together with the boundary conditions uniquely determine q_2. To do this we select formulation (A). Then from formulas (2.9.7b)

$$\begin{aligned} q_1^n = g_1^n &= A_1 f^{n+1/2} - B_1 f^n, & (x_1, x_2) &\in \gamma_2, \gamma_4; \\ q_2^n = g_2^n &= A_2 f^{n+1} - B_2 f^n, & (x_1, x_2) &\in \gamma_1, \gamma_3, \end{aligned} \tag{2.9.33}$$

q_2 is determined at points G from the equations

$$(E + \beta\,\tau\,\Lambda_2)\, q_1^n + (E - \alpha\,\tau\,\Lambda_1)\, q_2^n = 0, \tag{2.9.34}$$

which can be solved for q_2 since the operator $A_1 = E - \alpha\tau\Lambda_1$ can be inverted. Note that at points ω_2, ω_4 Eq. (2.9.34) takes the form

$$\frac{\beta\tau}{h_2^2} g_1^n + (E - \alpha\tau\Lambda_1)\, q_2^n = 0,$$

and, at all other points of G, the form

$$(E - \alpha\tau\Lambda_1)\, q_2^n = 0.$$

It is clear that an analogous method of introducing undetermined functions (right-hand sides q_1 and q_2) could be used for any scheme in fractional steps.

Consider now the boundary conditions on curved boundaries. As has been stated previously, in the case of the first boundary value problem, method (A) is equally applicable to a curved boundary. Considerably greater difficulties are encountered in the solution of Cauchy's problem with boundary conditions of the second kind. Then instead of Eq. (2.9.1a) we have (see Fig. 3)

$$\frac{\partial u}{\partial n} = \cos\varphi \frac{\partial u}{\partial x_1} + \sin\varphi \frac{\partial u}{\partial x_2} = f(x_1, x_2, t), \qquad (x_1, x_2, t) \in \Gamma. \tag{2.9.35}$$

The algorithm for the solution of boundary conditions (2.9.35) is given below. Let u^n be a difference solution of Cauchy's problem for Eq. (2.9.1) with boundary conditions (2.9.35), defined in region $\bar{G}$ at time $t = n\tau$. Determine u^{n+1} by using the values of $u^n(\gamma)$, already found. To do this we need to use the splitting scheme (2.9.2) with the boundary conditions

$$u^{n+1}\big|_\gamma = u^{n+1/2}\big|_\gamma = u^n(\gamma). \tag{2.9.36}$$

After this the boundary condition u^{n+1} is corrected with the use of Eq. (2.9.35). Denote by ω the set of the network points located close to γ, giving values of u^{n+1} in ω to γ. This can be done, for example, by projecting a normal from points ω to γ (see Fig. 3) and using the relation

$$u^{n-1}(x_\gamma) - u^{n+1}(x_\omega) = \Delta l\, f^{n+1}(x_{\bar{\omega}}), \tag{2.9.37}$$

where $f^{n+1}(x_\omega)$ is an approximation of $f(x_1, x_2, t)$ from Eq. (2.9.35). After this, Eq. (2.9.2) is solved again with boundary conditions from Eq. (2.9.36). After iterating, the next step is considered.

For the solution of boundary value problems of the second order in the case of equations with an unknown vector function, see Ch. 5.

Chapter 3

Application of the Method of Fractional Steps to Hyperbolic Equations

3.1 The simplest schemes for one-dimensional hyperbolic equations

Consider the equation of acoustics

$$\frac{\partial u}{\partial t} - a^2 \frac{\partial v}{\partial x} = 0; \qquad \frac{\partial v}{\partial t} - \frac{\partial u}{\partial x} = 0, \tag{3.1.1}$$

where u is the velocity; v is the specific volume; a is the velocity of sound, and x is the Lagrangian coordinate. Written in terms of Riemann invariants

$$r = u - a\,v; \qquad s = u + a\,v, \tag{3.1.2}$$

Eq. (3.1.1) takes the form

$$\frac{\partial r}{\partial t} + a \frac{\partial r}{\partial x} = 0; \qquad \frac{\partial s}{\partial t} - a \frac{\partial s}{\partial x} = 0. \tag{3.1.3}$$

We shall consider basic schemes for integrating Eqs. (3.1.1) or, equivalently Eqs. (3.1.3).

(a) *The scheme of running computation.* The explicit scheme of running computation has the form

$$\begin{gathered} \frac{\Delta_0 r}{\tau} + a \frac{\Delta_{-1} r}{h} = 0; \qquad \frac{\Delta_0 s}{\tau} - a \frac{\Delta_1 s}{h} = 0; \\ \Delta_0 = T_0 - E; \quad \Delta_1 = T_1 - E; \quad \Delta_{-1} = E - T_{-1}, \end{gathered} \tag{3.1.4}$$

where the shift operators T_1, T_{-1} are defined by relation (1.2.2), and the shift operator T_0 with respect to t is defined in an analogous way

$$T_0 f(x, t) = f(x, t + \tau). \tag{3.1.5}$$

Using suffix notation we have

$$\begin{gathered} \frac{r_i^{n+1} - r_i^n}{\tau} + a \frac{r_i^n - r_{i-1}^n}{h} = 0; \\ \frac{s_i^{n+1} - s_i^n}{\tau} - a \frac{s_{i+1}^n - s_i^n}{h} = 0. \end{gathered} \tag{3.1.6}$$

Scheme (3.1.4) in the variables u and v is

$$\begin{aligned} \frac{\Delta_0 u}{\tau} - a^2 \frac{\Delta_1 + \Delta_{-1}}{2h} v &= \frac{a\,h}{2} \frac{\Delta_1 \Delta_{-1}}{h^2} u; \\ \frac{\Delta_0 v}{\tau} - \frac{\Delta_1 + \Delta_{-1}}{2h} u &= \frac{a\,h}{2} \frac{\Delta_1 \Delta_{-1}}{h^2} v. \end{aligned} \tag{3.1.7}$$

The step operator of scheme (3.1.4) can easily be shown to be

$$\begin{aligned} r^{n+1} &= C\, r^n, \qquad C = E - \frac{a\tau}{h} \Delta_{-1}; \\ s^{n+1} &= D\, s^n, \qquad D = E + \frac{a\tau}{h} \Delta_1 \end{aligned} \tag{3.1.8}$$

and similarly, that of Eq. (3.1.7) is

$$\begin{aligned} u^{n+1} &= \frac{C+D}{2} u^n - a \frac{C-D}{2} v^n; \\ v^{n+1} &= -\frac{C-D}{2a} u^n + \frac{C+D}{2} v^n. \end{aligned} \tag{3.1.9}$$

Consider implicit schemes. The weighted implicit scheme is

$$\left.\begin{aligned} &\left[\frac{\Delta_0}{\tau} (\alpha_1 T_{-1} + \beta_1 E) + a \frac{\Delta_{-1}}{h} (\alpha_2 T_0 + \beta_2 E)\right] r = 0; \\ &\left[\frac{\Delta_0}{\tau} (\alpha_1 T_1 + \beta_1 E) - a \frac{\Delta_1}{h} (\alpha_2 T_0 + \beta_2 E)\right] s = 0; \\ &\alpha_s \geq 0; \quad \beta_s \geq 0; \quad \alpha_s + \beta_s = 1; \quad s = 1, 2. \end{aligned}\right\} \tag{3.1.10}$$

Using the index notation we have

$$\begin{aligned} &\left(\alpha_1 \frac{r_{i-1}^{n+1} - r_{i-1}^n}{\tau} + \beta_1 \frac{r_i^{n+1} - r_i^n}{\tau}\right) + \\ &\qquad + a\left(\alpha_2 \frac{r_i^{n+1} - r_{i-1}^{n+1}}{h} + \beta_2 \frac{r_i^n - r_{i-1}^n}{h}\right) = 0; \\ &\left(\alpha_1 \frac{s_{i+1}^{n+1} - s_{i+1}^n}{\tau} + \beta_1 \frac{s_i^{n+1} - s_i^n}{\tau}\right) - \\ &\qquad - a\left(\alpha_2 \frac{s_{i+1}^{n+1} - s_i^{n+1}}{h} + \beta_2 \frac{s_{i+1}^n - s_i^n}{h}\right) = 0. \end{aligned} \tag{3.1.11}$$

The step operator of scheme (3.1.10) is

$$r^{n+1} = C\, r^n; \qquad s^{n+1} = D\, s^n, \tag{3.1.12}$$

where

$$\begin{aligned} C &= \left(\alpha_1 T_{-1} + \beta_1 E + a\alpha_2 \tau \frac{\Delta_{-1}}{h}\right)^{-1} \left(\alpha_1 T_{-1} + \beta_1 E - a\beta_2 \tau \frac{\Delta_{-1}}{h}\right); \\ D &= \left(\alpha_1 T_1 + \beta_1 E - a\alpha_2 \tau \frac{\Delta_1}{h}\right)^{-1} \left(\alpha_1 T_1 + \beta_1 E + a\beta_2 \tau \frac{\Delta_1}{h}\right). \end{aligned} \tag{3.1.13}$$

In terms of the variables u and v we again use the formula (3.1.9), but with C and D given by Eq. (3.1.13).

The weighted scheme (3.1.10) gives us the known formulas for running computation, including the explicit majorant scheme ($\alpha_1 = 0$, $\beta_1 = 1$; $\alpha_2 = 0$, $\beta_2 = 1$); the implicit majorant scheme ($\alpha_1 = 0, \beta_1 = 1$; $\alpha_2 = 1, \beta_2 = 0$); and the scheme of second order accuracy ($\alpha_s = \beta_s = 1/2$, $s = 1, 2$)[1].

(b) *The "cross" scheme*

$$\frac{u^{n+1} - u^n}{\tau} - a^2 \frac{\Delta_{-1} v^n}{h} = 0; \qquad \frac{v^{n+1} - v^n}{\tau} - \frac{\Delta_1 u^{n+1}}{h} = 0. \tag{3.1.14}$$

(c) *The weighted implicit scheme*

$$\begin{aligned} \frac{u^{n+1} - u^n}{\tau} - a^2 \frac{\Delta_{-1}}{h} (\alpha_1 v^{n+1} + \beta_1 v^n) &= 0, \\ \frac{v^{n+1} - v^n}{\tau} - \frac{\Delta_1}{h} (\alpha_2 u^{n+1} + \beta_2 u^n) &= 0. \end{aligned} \tag{3.1.15}$$

With $\alpha_s = \beta_s = 1/2$ ($s = 1, 2$) Eq. (3.1.15) is of second order accuracy (the Neumann scheme); with $\alpha_1 = 0$, $\beta_1 = 1$, $\alpha_2 = 1$, $\beta_2 = 0$ it is transformed into the "cross" scheme.

The schemes (3.1.9), (3.1.14), and (3.1.15) are used as the basis for the construction of simple schemes of integration of equations of acoustics and of the wave equation in several dimensions.

3.2 Uniform implicit schemes for equations of hyperbolic type

The implicit approximation, as was shown for the first time by Laasonen [29], is naturally suited to parabolic equations with an infinite domain of dependence because in this case the domain of dependence is also infinite for the difference equation.

As opposed to those of parabolic type, hyperbolic equations have a finite domain of dependence; therefore, it would seem to be appropriate to use an explicit approximation. However, this is not so.

As shown by the well known Courant criterion, the stability requirements are determined by the numerical values at a given point. At the same time the accuracy requirements are determined by gradients. In the case of a flow with small gradients (flow in a river, atmospheric flows, etc.) the step τ, which is determined by accuracy requirements, is much larger than the step determined by stability requirements. Therefore the use of implicit schemes is also necessary for hyperbolic equations.

[1] Majorant schemes are those with positive coefficients (see [28]). In the case of constant coefficients these formulas, as a rule, satisfy the extremum property and are stable in C.

The implicit scheme for the equations of hydrodynamics, based on the method of running computation, was first suggested by L. D. Landau, N. N. Meiman and I. M. Khalatnikov [30]. O. A. Ladyzhenskaya [31] laid the first theoretical foundation for implicit schemes to be applied to the wave equation with variable coefficients. The application of implicit schemes in hydrodynamics is an accepted fact at the present time [32–35].

3.3 Implicit schemes for hyperbolic equations in several dimensions

The method of splitting for hyperbolic systems in several dimensions was suggested for the first time by K. A. Bagrinovskii and S. K. Godunov [36]. They considered only the explicit approximation, for which the splitting scheme does not offer many advantages over the usual explicit schemes.

In another work by N. N. Anuchina and the author [37] an implicit scheme was proposed for hyperbolic systems in several dimensions.

Let us consider the equation of acoustics in the two-dimensional case

$$\frac{\partial u_1}{\partial t} - a^2 \frac{\partial v}{\partial x_1} = 0, \quad \frac{\partial u_2}{\partial t} - a^2 \frac{\partial v}{\partial x_2} = 0, \quad \frac{\partial v}{\partial t} - \left(\frac{\partial u_1}{\partial x_1} + \frac{\partial u_2}{\partial x_2}\right) = 0. \qquad (3.3.1)$$

Eq. (3.3.1) in matrix form is

$$\frac{\partial f}{\partial t} = A f, \qquad (3.3.2)$$

where

$$A = \begin{Vmatrix} 0 & 0 & a^2 D_1 \\ 0 & 0 & a^2 D_2 \\ D_1 & D_2 & 0 \end{Vmatrix}, \quad D_i = \frac{\partial}{\partial x_i}, \quad f = \{u_1, u_2, v\}. \qquad (3.3.3)$$

Let us construct a splitting scheme based on the one-dimensional implicit scheme (3.1.15). At the first fractional step the following scheme is obtained, when gradients with respect to x_2 are neglected:

$$\frac{u_1^{n+1/2} - u_1^n}{\tau} - a^2 \frac{\Delta_{-1}}{h_1} (\alpha v^{n+1/2} + \beta v^n) = 0; \quad \frac{u_2^{n+1/2} - u_2^n}{\tau} = 0;$$
$$\frac{v^{n+1/2} - v^n}{\tau} - \frac{\Delta_1}{h_1} (\alpha u_1^{n+1/2} + \beta u_1^n) = 0. \qquad (3.3.4)$$

At the second fractional step, omitting gradients with respect to x_1, we have

$$\frac{u_1^{n+1} - u_1^{n+1/2}}{\tau} = 0; \quad \frac{u_2^{n+1} - u_2^{n+1/2}}{\tau} - a^2 \frac{\Delta_{-2}}{h_2} (\alpha v^{n+1} + \beta v^{n+1/2}) = 0;$$
$$\frac{v^{n+1} - v^{n+1/2}}{\tau} - \frac{\Delta_2}{h_2} (\alpha u_2^{n+1} + \beta u_2^{n+1/2}) = 0. \qquad (3.3.5)$$

For simplicity assume that

$$\alpha_1 = \alpha_2 = \alpha, \quad \beta_1 = \beta_2 = \beta.$$

Clearly the splitting scheme (3.3.4), (3.3.5) is absolutely stable for $\alpha \geq 1/2$. Let us show now that this scheme approximates system (3.3.1) or, what amounts to the same, the following equation, which is equivalent to Eq. (3.3.1)

$$\frac{\partial}{\partial t}\left[\frac{\partial^2 f}{\partial t^2} - a^2\left(\frac{\partial^2 f}{\partial x_1^2} + \frac{\partial^2 f}{\partial x_2^2}\right)\right] = 0, \tag{3.3.6}$$

which is satisfied by each of the values of u_1, u_2 and v. The approximation is proved by the method of elimination. By eliminating $u_1^{n+1/2}$ and $u_2^{n+1/2}$ from Eqs. (3.3.4) and (3.3.5) four operator equations are obtained with four unknowns, u_1^n, u_2^n, v^n, $v^{n+1/2}$ which are, in matrix form:

$$\left\|\begin{matrix} \frac{T_0 - E}{\tau} & 0 & -\beta a^2 \frac{\Delta_{-1}}{h_1} & -\alpha a^2 \frac{\Delta_{-1}}{h_1} \\ -\frac{\Delta_1}{h_1}(\alpha T_0 + \beta E) & 0 & -\frac{1}{\tau} E & \frac{1}{\tau} E \\ 0 & \frac{T_0 - E}{\tau} & -\alpha a^2 \frac{\Delta_{-2}}{h_2} T_0 & -\beta a^2 \frac{\Delta_{-2}}{h_2} \\ 0 & -\frac{\Delta_2}{h_2}(\alpha T_0 + \beta E) & \frac{1}{\tau} T_0 & -\frac{1}{\tau} E \end{matrix}\right\| \times$$

$$\times \left\|\begin{matrix} u_1^n \\ u_2^n \\ v^n \\ v^{n+1/2} \end{matrix}\right\| = 0. \tag{3.3.7}$$

Eliminating $v^{n+1/2}$ from Eq. (3.3.7) we find

$$\left.\begin{aligned} &\left[\frac{T_0 - E}{\tau} - \alpha \tau a^2 \frac{\Delta_1 \Delta_{-1}}{h_1^2}(\alpha T_0 + \beta E)\right] u_1^n - a^2 \frac{\Delta_{-1}}{h_1} v^n = 0; \\ &\left[\frac{T_0 - E}{\tau} + \beta \tau a^2 \frac{\Delta_2 \Delta_{-2}}{h_2^2}(\alpha T_0 + \beta E)\right] u_2^n - a^2 \frac{\Delta_{-2}}{h_2} T_0 v^n = 0; \\ &\frac{T_0 - E}{\tau} v^n - \frac{\Delta_1}{h_1}(\alpha T_0 + \beta E) u_1^n - \frac{\Delta_2}{h_2}(\alpha T_0 + \beta E) u_2^n = 0, \end{aligned}\right\} \tag{3.3.8}$$

which can also be expressed in matrix form

$$\frac{T_0 - E}{\tau} f^n = B f^n; \quad f^n = \{u_1^n, u_2^n, v^n\}, \tag{3.3.9}$$

where

$$B = \left\|\begin{matrix} \alpha \tau a^2 \frac{\Delta_1 \Delta_{-1}}{h_1^2}(\alpha T_0 + \beta E) & 0 & a^2 \frac{\Delta_{-1}}{h_1} \\ 0 & -\beta \tau a^2 \frac{\Delta_2 \Delta_{-2}}{h_2^2}(\alpha T_0 + \beta E) & a^2 \frac{\Delta_{-2}}{h_2} T_0 \\ \frac{\Delta_1}{h_1}(\alpha T_0 + \beta E) & \frac{\Delta_2}{h_2}(\alpha T_0 + \beta E) & 0 \end{matrix}\right\|. \tag{3.3.10}$$

It is easy to see that scheme (3.3.9) is consistent with scheme (3.3.2). Consistency can also be proved in another way. If the determinant of the operator scheme (3.3.7) is equated to zero, the following operator equation of the third order for any of the values $f^n = u_1^n, u_2^n, v^n, v^{n+1/2}$ is obtained:

$$\Big[\Big(\frac{T_0 - E}{\tau}\Big)^3 - a^2 \frac{T_0 - E}{\tau}(\alpha T_0 + \beta E)^2 \frac{\Delta_1 \Delta_{-1}}{h_1^2} - - a^2 \frac{T_0 - E}{\tau}(\alpha T_0 + \beta E)^2 \frac{\Delta_2 \Delta_{-2}}{h_2^2} + + a^4 \tau (\alpha^2 T_0 - \beta^2 E)(\alpha T_0 + \beta E)^2 \frac{\Delta_1 \Delta_{-1}}{h_1^2} \frac{\Delta_2 \Delta_{-2}}{h_2^2}\Big] f^n = 0. \qquad (3.3.11)$$

The scheme (3.3.11) is consistent with the equation

$$\frac{\partial^3 f}{\partial t^3} = a^2 \frac{\partial}{\partial t}\Big(\frac{\partial^2 f}{\partial x_1^2} + \frac{\partial^2 f}{\partial x_2^2}\Big) + \tau L f, \qquad (3.3.12)$$

where L is some operator, i.e., it is consistent with the equation

$$\frac{\partial}{\partial t}\Big[\frac{\partial^2 f}{\partial t^2} - a^2\Big(\frac{\partial^2 f}{\partial x_1^2} + \frac{\partial^2 f}{\partial x_2^2}\Big)\Big] = 0. \qquad (3.3.13)$$

Let $\alpha = \beta = 1/2$. Then Eq. (3.3.11) takes the form

$$(T_0 - E)\Big[\Big(\frac{T_0 - E}{\tau}\Big)^2 - \frac{a^2}{4}(T_0 + E)^2 \frac{\Delta_1 \Delta_{-1}}{h_1^2} - \frac{a^2}{4}(T_0 + E)^2 \frac{\Delta_2 \Delta_{-2}}{h_2^2} + + a^4 \tau^2 \frac{1}{16}(T_0 + E)^2 \frac{\Delta_1 \Delta_{-1}}{h_1^2} \frac{\Delta_2 \Delta_{-2}}{h_2^2}\Big] f^n = 0. \qquad (3.3.14)$$

After simplification by $(T_0 - E)$ we obtain the equation of the second order

$$\frac{f^{n+1} - 2f^n + f^{n-1}}{\tau^2} = a^2\Big(\frac{\Delta_1 \Delta_{-1}}{h_1^2} + \frac{\Delta_2 \Delta_{-2}}{h_2^2}\Big)\frac{f^{n-1} + 2f^n + f^{n+1}}{4} - - \frac{a^4}{16}\tau^2 \frac{\Delta_1 \Delta_{-1} \Delta_2 \Delta_{-2}}{h_1^2 h_2^2}(f^{n-1} + 2f^n + f^{n+1}), \qquad (3.3.15)$$

which has accuracy of order $O(\tau^2 + h^2)$.

Let us show that the scheme (3.3.15) can be carried out with the use of simple three-point sweeps. By taking v^n (specific volume) as f^n, the following expression is obtained from the first two equations of (3.3.7), after eliminating u_1^n:

$$\frac{v^{n+3/2} - v^{n+1} - v^{n+1/2} + v^n}{\tau^2} = a^2 \frac{\Delta_1 \Delta_{-1}}{h_1^2}(\alpha^2 v^{n+3/2} + \alpha\beta v^{n+1} + + \alpha\beta v^{n+1/2} + \beta^2 v^n). \qquad (3.3.16)$$

From the last two equations of (3.3.7) we obtain, after elimination of u_2^n,

$$\frac{v^{n+2} - v^{n+3/2} - v^{n+1} + v^{n+1/2}}{\tau^2} = a^2 \frac{\Delta_2 \Delta_{-2}}{h_2^2}(\alpha^2 v^{n+2} + \alpha\beta v^{n+3/2} + + \alpha\beta v^{n+1} + \beta^2 v^{n+1/2}). \qquad (3.3.17)$$

Eqs. (3.3.16) and (3.3.17) represent Eq. (3.3.15) for $\alpha = \beta = 1/2$.

3.4 The splitting scheme of running computation

By applying the running computation schemes (3.1.9) and (3.1.13) to system (3.3.1) at each fractional step the following expression is obtained

$$f^{n+1/2} = \sigma_1 f^n; \qquad f^{n+1} = \sigma_2 f^{n+1/2}, \tag{3.4.1}$$

where

$$\sigma_1 = \begin{Vmatrix} \frac{C_1 + D_1}{2} & 0 & -a\frac{C_1 - D_1}{2} \\ 0 & E & 0 \\ -\frac{C_1 - D_1}{2a} & 0 & \frac{C_1 + D_1}{2} \end{Vmatrix}, \qquad \sigma_2 = \begin{Vmatrix} E & 0 & 0 \\ 0 & \frac{C_2 + D_2}{2} & -a\frac{C_2 - D_2}{2} \\ 0 & -\frac{C_2 - D_2}{2a} & \frac{C_2 + D_2}{2} \end{Vmatrix}. \tag{3.4.2}$$

Operators C_s and D_s, according to (3.1.13), are expressed by

$$\left.\begin{aligned} C_s &= \left(\alpha_1 T_{-s} + \beta_1 E + \alpha_2 a\tau\frac{\Delta_{-s}}{h_s}\right)^{-1}\left(\alpha_1 T_{-s} + \beta_1 E - \beta_2 a\tau\frac{\Delta_{-s}}{h_s}\right); \\ D_s &= \left(\alpha_1 T_s + \beta_1 E - \alpha_2 a\tau\frac{\Delta_s}{h_s}\right)^{-1}\left(\alpha_1 T_s + \beta_1 E + \beta_2 a\tau\frac{\Delta_s}{h_s}\right); \end{aligned}\right\} \tag{3.4.3}$$

$$s = 1, 2.$$

The step operator σ of the scheme in whole steps,

$$f^{n+1} = \sigma f^n \tag{3.4.4}$$

is expressed by the product of matrices σ_2, σ_1

$$\sigma = \sigma_2\sigma_1 = \begin{Vmatrix} \frac{C_1+D_1}{2} & 0 & -a\frac{C_1-D_1}{2} \\ \frac{C_2-D_2}{2}\cdot\frac{C_1-D_1}{2} & \frac{C_2+D_2}{2} & -a\frac{C_2-D_2}{2}\cdot\frac{C_1+D_1}{2} \\ -\frac{(C_2+D_2)(C_1-D_1)}{4a} & -\frac{C_2-D_2}{2a} & \frac{C_2+D_2}{2}\cdot\frac{C_1+D_1}{2} \end{Vmatrix}. \tag{3.4.5}$$

The matrix

$$\frac{\sigma - E}{\tau} = \begin{Vmatrix} \frac{\frac{C_1+D_1}{2} - E}{\tau} & 0 & -a\frac{C_1-D_1}{2\tau} \\ \frac{(C_2-D_2)(C_1-D_1)}{4\tau} & \frac{\frac{C_2+D_2}{2} - E}{\tau} & -a\frac{(C_2-D_2)(C_1+D_1)}{4\tau} \\ -\frac{(C_2+D_2)(C_1-D_1)}{4a\tau} & -\frac{C_2-D_2}{2a\tau} & \frac{\frac{(C_2+D_2)(C_1+D_1)}{4} - E}{\tau} \end{Vmatrix} \tag{3.4.6}$$

approximates the matrix A of Eq. (3.3.3). It is easy to show that

$$\frac{C_i + D_i}{2} \sim E; \quad \frac{\frac{C_i + D_i}{2} - E}{\tau} \sim 0; \quad \frac{C_i - D_i}{2a\tau} \sim -\frac{\Delta_i}{h_i}. \tag{3.4.7}$$

It follows from this that

$$\frac{\sigma - E}{\tau} \sim A. \tag{3.4.8}$$

Thus we obtained the required result.

Because each scheme of running computation is stable:

$$\|\sigma_1\| \le 1; \quad \|\sigma_2\| \le 1, \tag{3.4.8'}$$

the following inequality is valid:

$$\|\sigma\| \le 1. \tag{3.4.9}$$

Consequently the scheme of splitting for running computation is reduced to scheme (3.3.2). The following two remarks should be made.

(i) Despite the fact that the one-dimensional scheme of running computation could have second order accuracy (at $\alpha = \beta = 1/2$), the scheme of splitting in whole steps is unsymmetrical and therefore is not sufficiently accurate. In order to increase the accuracy scheme (3.4.4) should be made symmetrical by assuming that (see [38] and also [94]):

$$\Sigma = \frac{\sigma + \sigma^*}{2} = \frac{\sigma_2\,\sigma_1 + \sigma_1\,\sigma_2}{2}, \tag{3.4.10}$$

$$\sigma^* = \sigma_1\sigma_2 = \left\| \begin{matrix} \frac{C_1 + D_1}{2} & \frac{(C_1 - D_1)(C_2 - D_2)}{4} & -a\frac{(C_1 - D_1)(C_2 + D_2)}{4} \\ 0 & \frac{C_2 + D_2}{2} & -a\frac{C_2 - D_2}{2} \\ -\frac{C_1 - D_1}{2a} & -\frac{(C_1 + D_1)(C_2 - D_2)}{4a} & \frac{(C_1 + D_1)(C_2 + D_2)}{4} \end{matrix} \right\|. \tag{3.4.11}$$

Then the step operator Σ has the symmetrical form

$$\Sigma = \left\| \begin{matrix} \frac{C_1 + D_1}{2} & \frac{(C_1 - D_1)(C_2 - D_2)}{4 \cdot 2} & -a\frac{(C_1 - D_1)(C_2 + D_2 + 2E)}{4 \cdot 2} \\ \frac{(C_2 - D_2)(C_1 - D_1)}{4 \cdot 2} & \frac{C_2 + D_2}{2} & -\frac{a(C_2 - D_2)(C_1 + D_1 + 2E)}{4 \cdot 2} \\ -\frac{C_2 + D_2 + 2E}{2} \cdot \frac{C_1 - D_1}{4a} & -\frac{C_1 + D_1 + 2E}{2} \cdot \frac{C_2 - D_2}{4a} & \frac{(C_2 + D_2)(C_1 + D_1) + (C_1 + D_1)(C_2 + D_2)}{4 \cdot 2} \end{matrix} \right\|. \tag{3.4.12}$$

(ii) During the multiplication of matrix operators their elements (operators) should be multiplied, taking into account their possible non-commutative property.

3.5 Method of approximate factorization for the wave equation

In the preceding sections schemes in fractional steps for the equation of acoustics (3.3.1) were considered. As was noted, the values u_1, u_2, and v satisfy Eq. (3.3.6). The value v satisfies the wave equation

$$\frac{\partial^2 v}{\partial t^2} - a^2 \left(\frac{\partial^2 v}{\partial x_1^2} + \frac{\partial^2 v}{\partial x_2^2} \right) = 0. \tag{3.5.1}$$

Therefore the four-layer integration scheme (3.3.11), or the equivalent scheme in fractional steps (3.3.16), (3.3.17), can be simplified when applied to v. E. G. D'yakonov [101] suggested the three-layer integration scheme of Eq. (3.5.1) based on approximate factorization of the operator.

Let

$$\begin{gathered} \frac{v^{n+1} - 2v^n + v^{n-1}}{\tau^2} = \Lambda \frac{v^{n+1} + v^{n-1}}{2}; \\ \Lambda = \Lambda_1 + \Lambda_2, \quad \Lambda_s = a^2 \frac{\Delta_s \Delta_{-s}}{h_s^2}, \quad s = 1, 2, \end{gathered} \tag{3.5.2}$$

be a uniform scheme of second order accuracy consistent with Eq. (3.5.1). Rewrite Eq. (3.5.2) in the form

$$\left(E - \frac{1}{2} \tau^2 \Lambda \right) \frac{v^{n+1} + v^{n-1}}{2} = v^n. \tag{3.5.3}$$

Let us factorize approximately the operator on the left-hand side of Eq. (3.5.3)

$$E - \frac{1}{2} \tau^2 \Lambda \sim \left(E - \frac{1}{2} \tau^2 \Lambda_1 \right) \left(E - \frac{1}{2} \tau^2 \Lambda_2 \right) \tag{3.5.4}$$

and replace the scheme (3.5.3) by the factorized expression

$$\left(E - \frac{1}{2} \tau^2 \Lambda_1 \right) \left(E - \frac{1}{2} \tau^2 \Lambda_2 \right) \frac{v^{n+1} + v^{n-1}}{2} = v^n. \tag{3.5.5}$$

Scheme (3.5.5) is effected in two sweeps

$$\begin{gathered} \left(E - \frac{1}{2} \tau^2 \Lambda_1 \right) v^{n+1/2} = v^n; \\ \left(E - \frac{1}{2} \tau^2 \Lambda_2 \right) \frac{v^{n+1} + v^{n-1}}{2} = v^{n+1/2}. \end{gathered} \tag{3.5.6}$$

It is easy to see that scheme (3.5.5) is stable and is of second order accuracy.

Splitting schemes for the equation governing vibrations were constructed by A. N. Konovalov [40] and A. A. Samarski [41].

3.6 The method of splitting and majorant schemes

The schemes of running computation for one-dimensional hyperbolic systems, which are based on the majorant approximation of the equations in invariant form, lead to schemes with positive coefficients (majorant schemes).

K. O. Friedrichs [42] introduced the idea of positive schemes (schemes with positive matrices). The scheme

$$\left.\begin{aligned} u^{n+1}(x_1, \ldots, x_m) &= \sum_{\alpha} C_{\alpha_1 \ldots \alpha_m} u^n(x_1 + \alpha_1 h_1, \ldots, x_m + \alpha_m h_m) \\ &= \sum_{\alpha} C_{\alpha_1 \ldots \alpha_m} T_1^{\alpha_1} \ldots T_m^{\alpha_m} u^n(x_1, \ldots, x_m), \\ \alpha_i &= -q_i, \ldots, q_i, \quad i = 1, \ldots, m, \end{aligned}\right\} \tag{3.6.1}$$

is positive if all $C_{\alpha_1 \ldots \alpha_m}$ are positive[1].

As applied to hyperbolic systems in several dimensions the method of splitting makes it possible to obtain majorant approximations. Let

$$\frac{\partial u}{\partial t} + \sum_{\alpha=1}^{m} A_\alpha \frac{\partial u}{\partial x_\alpha} + B u = 0 \tag{3.6.2}$$

be a linear (symmetrical according to Friedrichs) hyperbolic system. Here $u = \{u_1, \ldots, u_p\}$ are vector-functions; A_α and B are symmetric $p \times p$ matrices.

The author [28] demonstrated the possibility of constructing majorant schemes by the method of splitting.

Following N. N. Anuchina [43] we show that it is possible to construct majorant approximations of type (3.6.1) for equations of type (3.6.2). If A_α are positive matrices, then applying the approximation

$$\frac{u^{n+1}(x) - u^n(x)}{\tau} + \sum_{\alpha=1}^{m} A_\alpha \frac{E - T_{-\alpha}}{h_\alpha} u^n(x) + B u^n(x) = 0, \tag{3.6.3}$$

the following scheme is obtained

$$u^{n+1}(x) = C_0 u^n(x) + \sum_{\alpha=1}^{m} C_{-\alpha} T_{-\alpha} u^n(x), \tag{3.6.4}$$

where

$$\left.\begin{aligned} C_0 &= E - \sum_{\alpha=1}^{m} \frac{\tau}{h_\alpha} A_\alpha - \tau B, \\ C_{-1} &= \frac{\tau}{h_1} A_1, \quad C_{-2} = \frac{\tau}{h_2} A_2, \quad \ldots, \quad C_{-m} = \frac{\tau}{h_m} A_m; \\ T_{-\alpha} f(x_1, \ldots, x_m) &= f(x_1, \ldots, x_\alpha - h_\alpha, \ldots, x_m). \end{aligned}\right\} \tag{3.6.5}$$

When τ/h is sufficiently small, the matrix C_0 is positive and the matrices $C_{-\alpha}$ are likewise positive.

If A_α are negative matrices then $E - T_{-\alpha}$ should be replaced by $T_\alpha - E$ in the scheme (3.6.3). The resulting scheme is

$$u^{n+1}(x) = C_0 u^n(x) + \sum_{\alpha=1}^{m} C_\alpha T_\alpha u^n(x), \tag{3.6.6}$$

---

[1] A square symmetric matrix A is called positive if all its eigenvalues are positive.

where

$$C_0 = E + \sum_{\alpha=1}^{m} \frac{\tau}{h_\alpha} A_\alpha - \tau B,$$

$$C_1 = -\frac{\tau}{h_1} A_1, \quad C_2 = -\frac{\tau}{h_2} A_2, \quad \ldots, \quad C_m = -\frac{\tau}{h_m} A_m;$$

$$T_\alpha f(x_1, \ldots, x_m) = f(x_1, \ldots, x_\alpha + h_\alpha, \ldots, x_m),$$

C_α are identically positive; C_0 is positive for sufficiently small τ/h.

If A_α are indefinite matrices then the following representation is possible

$$A_\alpha = \overset{1}{A}_\alpha + \overset{2}{A}_\alpha, \tag{3.6.7}$$

where the matrix $\overset{1}{A}_\alpha$ is non-negative, and the matrix $\overset{2}{A}_\alpha$ is non-positive. Then the scheme

$$\frac{u^{n+1}(x) - u^n(x)}{\tau} + \sum_{\alpha=1}^{m} \left(\overset{1}{A}_\alpha \frac{E - T_{-\alpha}}{h_\alpha} + \overset{2}{A}_\alpha \frac{T_\alpha - E}{h_\alpha} \right) u^n(x) + B\, u^n(x) = 0 \tag{3.6.8}$$

is reduced to

$$u^{n+1}(x) = C_0\, u^n(x) + \sum_{\alpha=1}^{m} C_{-\alpha} T_{-\alpha} u^n(x) + \sum_{\alpha=1}^{m} C_\alpha T_\alpha u^n(x), \tag{3.6.9}$$

where

$$\left.\begin{aligned} C_0 &= E - \sum_{\alpha=1}^{m} \frac{\tau}{h_\alpha} \overset{1}{A}_\alpha + \sum_{\alpha=1}^{m} \frac{\tau}{h_\alpha} \overset{2}{A}_\alpha - \tau B; \\ C_{-\alpha} &= \frac{\tau}{h_\alpha} \overset{1}{A}_\alpha; \quad C_\alpha = -\frac{\tau}{h_\alpha} \overset{2}{A}_\alpha, \quad \alpha = 1, \ldots, m. \end{aligned}\right. \tag{3.6.10}$$

Matrices $C_{-\alpha}$, C_α are always positive and the matrix C_0 is positive for sufficiently small τ/h.

The method of splitting makes it possible to construct easily realizable positive schemes. Consider the explicit majorant scheme of splitting

$$\frac{u^{n+s/m} - u^{n+(s-1)/m}}{\tau} + \left(\overset{1}{A}_s \frac{E - T_{-s}}{h_s} + \overset{2}{A}_s \frac{T_s - E}{h_s} + \frac{B}{m} \right) u^{n+(s-1)/m} = 0,$$

$$s = 1, \ldots, m. \tag{3.6.11}$$

The scheme (3.6.11) is reduced to

$$u^{n+s/m} = C_s\, u^{n+(s-1)/m}, \tag{3.6.12}$$

where

$$\left.\begin{aligned} C_s &= C_{0s} + C_{-s} T_{-s} + C_s T_s; \\ C_{0s} &= E - \frac{\tau}{h_s} \overset{1}{A}_s + \frac{\tau}{h_s} \overset{2}{A}_s - \frac{\tau B}{m}; \\ C_{-s} &= \frac{\tau}{h_s} \overset{1}{A}_s; \quad C_s = -\frac{\tau}{h_s} \overset{2}{A}_s, \quad s = 1, \ldots, m. \end{aligned}\right\} \tag{3.6.13}$$

The operators C_{-s} and C_s are always positive, and operators C_{0s} are positive for sufficiently small τ/h. Consequently the scheme

$$u^{n+1}(x) = C_m C_{m-1} \ldots C_1 u^n(x) \tag{3.6.14}$$

is positive.

Majorant schemes occupy a special place in difference methods. Despite the fact that they are of low accuracy, they are optimally stable at the same time. They produce good convergence in the space C and are simple in practical application.

Chapter 4

Application of the Method of Fractional Steps to Boundary Value Problems for Laplace's and Poisson's Equations

4.1 The relation between steady and unsteady problems

Consider the Dirichlet problem in the rectangular region G

$$\frac{\partial^2 u}{\partial x_1^2} + \frac{\partial^2 u}{\partial x_2^2} = 0, \tag{4.1.1}$$

$$u(x_1, x_2) = f(x_1, x_2), \qquad (x_1, x_2) \in \gamma, \tag{4.1.2}$$

where γ is the boundary G, $G = \{0 < x_i < \pi, i = 1, 2\}$. Along with Eq. (4.1.1) consider the unsteady problem

$$\frac{\partial u}{\partial t} = a^2 \left(\frac{\partial^2 u}{\partial x_1^2} + \frac{\partial^2 u}{\partial x_2^2}\right), \tag{4.1.3}$$

$$u(x_1, x_2, 0) = u_0(x_1, x_2) \tag{4.1.4}$$

with the same steady boundary conditions (4.1.2). Denote by $U(x_1, x_2)$ the solution of the problem (4.1.1), (4.1.2), and by $u(x_1, x_2, t)$ the solution of the problem (4.1.3), (4.1.4), (4.1.2). Then

$$v(x_1, x_2, t) = u(x_1, x_2, t) - U(x_1, x_2) \tag{4.1.5}$$

satisfies Eq. (4.1.3) with initial conditions

$$v(x_1, x_2, 0) = v_0(x_1, x_2) = u_0(x_1, x_2) - U(x_1, x_2) \tag{4.1.6}$$

and zero boundary conditions

$$v(x_1, x_2, t) = 0, \qquad (x_1, x_2) \in \gamma. \tag{4.1.7}$$

$v(x_1, x_2, t)$ is represented in the form

$$v(x_1, x_2, t) = \sum_{k_1=1}^{\infty} \sum_{k_2=1}^{\infty} A_{k_1 k_2}(t) \sin k_1 x_1 \sin k_2 x_2, \tag{4.1.8}$$

where

$$A_{k_1 k_2}(t) = a_{k_1 k_2} e^{-a^2 (k_1^2 + k_2^2) t} \tag{4.1.9}$$

is the Fourier coefficient of the function $v(x_1, x_2, t)$; $a_{k_1 k_2}$ is the Fourier coefficient of the function $v_0(x_1, x_2)$.

Formulas (4.1.8) and (4.1.9) can be represented in operator form

$$v = S(t)\, \dot{v}_0. \tag{4.1.10}$$

The operator $S(t)$ in space $L_2(G)$ has the norm

$$\| S(t) \| = e^{-2a^2 t}. \tag{4.1.11}$$

The step operator $S(\tau)$ has the norm

$$\| S(\tau) \| = e^{-2a^2 \tau}. \tag{4.1.12}$$

It follows from this that

$$\| S(t) \| \to 0, \qquad t \to \infty. \tag{4.1.13}$$

This means that

$$\| v(x_1, x_2, t) \| = \| u(x_1, x_2, t) - U(x_1, x_2) \| \to 0 \tag{4.1.14}$$

as $t \to \infty$, i.e., the solution of the unsteady problem approaches the solution of the steady problem with the same boundary conditions, regardless of the chcice of initial data.

It is evident that there exist many unsteady equations the solutions of which converge to the solution of the steady problem. Instead of Eq. (4.1.3) the following equation, for example, can be considered:

$$\frac{\partial u}{\partial t} + b^2 \frac{\partial^2 u}{\partial t^2} = a^2 \left(\frac{\partial^2 u}{\partial x_1^2} + \frac{\partial^2 u}{\partial x_2^2} \right). \tag{4.1.15}$$

The unsteady equation takes on the character of the equation of damped oscillations. It can be shown that in the case of Eq. (4.1.15) for any initial data related to $u(x_1, x_2, 0)$, $\partial u/\partial t|_{t=0}$ the solution of Eq. (4.1.15) converges to $U(x_1, x_2)$.

The unsteady equations (4.1.3) and (4.1.15) in these examples describe certain definite physical processes, namely, the diffusion of heat, and propagation of damped oscillations. However, in other cases the corresponding unsteady equations could have only a formal meaning, satisfying the mathematical conditions of damping (4.1.13) without describing any definite physical process.

This is true for the equation of elastic equilibrium

$$\Delta \Delta u = \frac{\partial^4 u}{\partial x_1^4} + 2 \frac{\partial^4 u}{\partial x_1^2\, \partial x_2^2} + \frac{\partial^4 u}{\partial x_2^4} = 0. \tag{4.1.16}$$

The unsteady equation

$$\frac{\partial u}{\partial t} + b^2 \frac{\partial^2 u}{\partial t^2} + a^2 \left[\frac{\partial^4 u}{\partial x_1^4} + 2 \frac{\partial^4 u}{\partial x_1^2\, \partial x_2^2} + \frac{\partial^4 u}{\partial x_2^4} \right] = 0 \tag{4.1.17}$$

has the solution operator $S(t)$ which satisfies the damping condition (4.1.13); and yet it is difficult to find a physical model which could be described by Eq. (4.1.17). This, however, will not prevent us from using Eq. (4.1.17) for the construction of iterative schemes.

4.2 The integration schemes of unsteady problems and iterative schemes

The relation between unsteady and steady equations demonstrated above can be transferred completely to difference schemes. Consider, for example, a two-layer difference scheme for Eq. (4.1.3),

$$\frac{u^{n+1}-u^n}{\tau_n} = \Omega_1 u^{n+1} + \Omega_2 u^n, \qquad \tau_n = a^2(t_{n+1}-t_n), \tag{4.2.1}$$

where Ω_1, Ω_2 are certain spatial difference operators which depend on τ_n, h_1, h_2.

Let u^n be the solution of Eq. (4.2.1) which satisfies a certain initial condition

$$u^0 = u_0 \tag{4.2.2}$$

and the steady boundary conditions

$$u^n = f, \qquad (x_1, x_2) \in \gamma. \tag{4.2.3}$$

Let w be the solution of the difference boundary problem

$$\begin{gathered} \Lambda w = 0, \qquad (x_1, x_2) \in G; \qquad w = f, \qquad (x_1, x_2) \in \gamma; \\ \Lambda = \Lambda_1 + \Lambda_2, \qquad \Lambda_i = \frac{\Delta_i \Delta_{-i}}{h_i^2}, \qquad i = 1, 2, \end{gathered} \tag{4.2.4}$$

where f is the function from Eq. (4.2.3).

The difference

$$v^n = u^n - w$$

satisfies the equation

$$\frac{v^{n+1}-v^n}{\tau_n} = \Omega_1 v^{n+1} + \Omega_2 v^n + (\Omega_1 + \Omega_2 - \Lambda) w.$$

Remember that the operators Ω_1, Ω_2 depend on τ_n, h_1, h_2; while the operators Λ, Λ_1, Λ_2 depend only on h_1, h_2.

Let us introduce the quantities

$$R_n = (\Omega_1 + \Omega_2 - \Lambda)_{n-1} w, \qquad R = \max \| R_n \|, \tag{4.2.5}$$

where $\| R_n \|$ is taken for all τ_n, bounded by a certain value $\tau_{\max}$.

Because of Eq. (1.2.24) we have

$$v^n = C_{n0} v^0 + \sum_{\alpha=1}^{n} C_{n\alpha} r_\alpha, \tag{4.2.6}$$

where

$$\begin{gathered} C_{n\alpha} = C_n C_{n-1} \ldots C_{\alpha+1}, \qquad C_n = (E - \tau_n \Omega_1)^{-1} (E + \tau_n \Omega_2), \\ r_\alpha = \tau_{\alpha-1} (E - \tau_{\alpha-1} \Omega_1)^{-1} R_\alpha. \end{gathered} \tag{4.2.7}$$

If the scheme (4.2.1) is strongly stable for a given law of approach to the limit (see Sec. 1.2), i.e.,

$$\| C_n \| \le 1 - k \tau_n, \tag{4.2.8}$$

where the constant $k > 0$ does not depend on τ_n, then the relation

$$\|v^n\| \leq (1 - k\,\tau_{\min})^n \|v^0\| + \tau_{\max} R \sum_{\alpha=1}^{n} (1 - k\,\tau_{\min})^{n-\alpha}$$

$$= (1 - k\,\tau_{\min})^n \|v^0\| + \frac{R}{k} \frac{\tau_{\max}}{\tau_{\min}} [1 - (1 - k\,\tau_{\min})^n], \tag{4.2.9}$$

is valid, where $\tau_{\max}$, $\tau_{\min}$ are the bounds of τ_n

$$\tau_{\min} \leq \tau_n \leq \tau_{\max}.$$

For fixed $\tau_{\min}$, $\tau_{\max}$ and as $n \to \infty$ we have

$$\lim \|v^n\| \leq \frac{R}{k} \frac{\tau_{\max}}{\tau_{\min}}. \tag{4.2.10}$$

Since the approximation $\Omega_1 + \Omega_2 \sim \Lambda$ is valid for scheme (4.2.1), then as $\tau_{\max} \to 0$, $h(\tau_{\max}) \to 0$, $R \to 0$, and from Eq. (4.2.10) it follows that

$$\lim \|v^n\| = 0$$

for

$$n \to \infty, \qquad \tau_{\max} \to 0, \qquad \frac{\tau_{\max}}{\tau_{\min}} = O(1).$$

Thus, with the approximation condition $\Omega_1 + \Omega_2 \sim \Lambda$ and strong stability of scheme (4.2.1), the solution u^n of the problem (4.2.1)–(4.2.3) approaches the solution w of problem (4.2.4) if $n \to \infty$, $\tau_{\max} \to 0$, $h(\tau_{\max}) \to 0$, $\tau_{\max}/\tau_{\min} = O(1)$, and u_0 is arbitrary. It follows from this that any strongly stable integration scheme (4.2.1) of Eq. (4.1.3) may be considered simultaneously as the iterative solution scheme of the boundary value problem (4.2.4). The integration step τ_n in this case can be regarded as an iterative parameter, or as a relaxation parameter.

As was shown before, the convergence $u^n \to w$ requires, generally speaking, a refinement of τ and h. If the approximation condition is required in stronger form

$$(\Omega_1 + \Omega_2 - \Lambda)\, w \to 0, \tag{4.2.11}$$

where $\tau \to 0$, and h is fixed, then the convergence $u^n \to w$ may occur for fixed h, but for very small τ_n, in general. In the case of still stronger approximation requirements the convergence $u^n \to w$ occurs for arbitrary τ and h. Suppose that the relation

$$\Omega_1 + \Omega_2 = \Lambda \tag{4.2.12}$$

is satisfied at any τ. The condition (4.2.12) is called the condition of complete consistency. If condition (4.2.12) is satisfied the convergence $u^n \to w$ occurs if and only if

$$\|C_{n0}\| \to 0, \qquad n \to \infty. \tag{4.2.13}$$

The condition (4.2.13) is the condition of asymptotic stability which, as is known, follows from the strong stability conditions (see Sec. 1.2). Condition (4.2.12) can be expressed in a more general form. Consider two-layer iterative schemes of the type

$$B\left(\frac{u^{n+1}-u^n}{\tau}\right) = \Lambda u^n, \tag{4.2.14}$$

where B is a certain linear operator. The expression (4.2.14) is the canonical form of two-layer iterative schemes and it is called the scheme of universal algorithm [87].

Let w be the solution of the stationary problem (4.2.4). Then $v^n = w - u^n$ satisfies Eq. (4.2.14)

$$B\left(\frac{v^{n+1}-v^n}{\tau}\right) = \Lambda v^n.$$

The step operator C for the scheme (4.2.14) is

$$C = E + \tau B^{-1} \Lambda. \tag{4.2.15}$$

If the operator C from Eq. (4.2.15) satisfies the strong stability condition, then we have convergence $v^n \to 0$, $u^n \to w$ for any τ and h which are located within the region of strong stability. Thus, we can consider the reduction of the iterative scheme to form (4.2.14) as a condition of complete consistency.

In the case when the B operator is a polynomial in τ and in a set of some finite space operators, it is easy to use Eq. (4.2.1) instead of Eq. (4.2.14) together with the complete consistency condition (4.2.12). In the case of a B operator of more complex structure such a transformation is impossible. Later, we shall consider the possibility of putting a scheme in the form (4.2.14) as a definition of complete consistency.

For two-layer schemes not satisfying the conditions of complete consistency we have

$$B\left(\frac{u^{n+1}-u^n}{\tau}\right) = \Omega u^n, \tag{4.2.16}$$

where the Ω operator approximates the Λ operator for some passage to the limit law $\tau \to 0$, $h \to 0$.

Let us determine now the speed of convergence of an iterative scheme with complete consistency and strong stability for fixed parameters τ, h_1, h_2. When h_1, h_2 are fixed, τ is selected in an optimal way so that the norm of the C operator becomes minimal. Let

$$\min_{\tau} \| C(\tau, h_1, h_2) \| = 1 - \varepsilon(h_1, h_2). \tag{4.2.17}$$

The speed of convergence can be estimated asymptotically by letting h_1, h_2 approach zero and by expanding $\varepsilon(h_1, h_2)$ in series in h_1 and h_2.

Assume, for simplicity, that $h_1 = h_2 = h$ and that the asymptotic evaluation

$$\min_{\tau} \| C(\tau, h_1, h_2) \| = 1 - R h^{\alpha}, \quad R > 0, \quad \alpha > 0 \tag{4.2.18}$$

is valid. It is not difficult to show that, in order to decrease the norm of the deviation v^n by $(q = 1/\varepsilon)$ times, we need m iterations, where

$$m \sim \frac{|\ln \varepsilon|}{R h^{\alpha}} \sim \frac{\ln q}{R} N^{\alpha}, \quad N \sim \frac{1}{h}. \tag{4.2.19}$$

A study of the stability and the speed of convergence is especially easy, for schemes with complete consistency, in the case when the C operator has functions (harmonics) $\sin k_1 x_1 \sin k_2 x_2$ as its eigen-functions. In this case we have

$$\left.\begin{aligned} v^n &= \sum_{k_1, k_2 = 1}^{N_1, N_2} a^n(\tau, h_1, h_2, k_1, k_2) \sin k_1 x_1 \sin k_2 x_2; \\ a^n &= \varrho(\tau_n, h_1, h_2, k_1, k_2)\, a^{n-1} = \varrho^n a^0; \\ \varrho^n &= \varrho(\tau_1, h_1, h_2, k_1, k_2) \ldots \varrho(\tau_n, h_1, h_2, k_1, k_2); \\ \| C^n \| &= \max_{k_1, k_2} |\varrho(\tau_n, h_1, h_2, k_1, k_2)|, \end{aligned}\right\} \tag{4.2.20}$$

where C_n is the transition operator from the $(n - 1)$-th to n-th iteration. Thus, the norm of the iterative step operator is defined as the maximum of the modulus of the amplification factor ϱ, corresponding to the harmonics $\sin k_1 x_1 \sin k_2 x_2$.

For operators, the eigen-functions of which are not harmonics, the harmonic analysis of stability has only a heuristic character.

4.3 Iterative schemes for Laplace's equation in two dimensions

We now give comparative analysis of the iterative schemes for the Laplace equation.

We need here the following notation: $\varrho = \varrho(\tau, h_1, h_2, k_1, k_2)$ is the increment coefficient of the harmonics $\sin k_1 x_1 \sin k_2 x_2$ following one iteration with a step τ;

$$\left.\begin{aligned} \varrho_0 &= \| C(\tau, h_1, h_2) \| = \max_{k_1, k_2} |\varrho|; \\ \varrho_1 &= \min_{\tau} \| C(\tau, h_1, h_2) \| = \min_{\tau} \varrho_0; \\ \varrho_n &= \varrho(\tau_1, h_1, h_2, k_1, k_2)\, \varrho(\tau_2, h_1, h_2, k_1, k_2) \ldots \varrho(\tau_n, h_1, h_2, k_1, k_2), \end{aligned}\right\} \tag{4.3.1}$$

ϱ_n is the increment coefficient of the harmonics $\sin k_1 x_1 \sin k_2 x_2$ after n iterations with steps $\tau_1, \ldots, \tau_n$,

$$a_i = \frac{4 a^2 \tau}{h_i^2} \sin^2 \frac{k_i h_i}{2}, \quad r_i = \frac{a^2 \tau}{h_i^2}, \quad i = 1, 2,$$

m is the number of iterations necessary to achieve a given accuracy ε; $k_i = 1, 2, \ldots, N_i$, $(N_i + 1) h_i = \pi$, $i = 1, 2$. In what follows asymptotic estimates are always made and therefore it is assumed that

$$\sin \frac{h_i}{2} \simeq \frac{h_i}{2}, \qquad \cos \frac{h_i}{2} \simeq 1 - \frac{h_i^2}{8}$$

approximately. In addition, for simplicity, we assume that $h_1 = h_2 = h$; $a^2 = 1$.

(i) *The explicit scheme*

$$\frac{u^{n+1} - u^n}{\tau} = \Lambda u^n, \qquad \Lambda = \Lambda_1 + \Lambda_2, \qquad \Lambda_i = \frac{\Delta_i \Delta_{-i}}{h_i^2}, \tag{4.3.2}$$

$$\varrho = 1 - (a_1 + a_2),$$
$$\varrho_0 = \max \left\{ \left|1 - 2\tau\right|, \left|1 - 8r\left(1 - \frac{h^2}{4}\right)\right| \right\}. \tag{4.3.3}$$

Complete consistency is always achieved, and strong stability applies under the conditions $r \leq 1/4$. It follows from this that

$$\varrho_1 = 1 - \frac{1}{2} h^2, \tag{4.3.4}$$

$$m \simeq 2 \left|\ln \varepsilon\right| N^2. \tag{4.3.5}$$

(ii) *The relaxation scheme along a line*

$$\frac{u^{n+1} - u^n}{\tau} = \Lambda_1 u^{n+1} + \Lambda_2 u^n. \tag{4.3.6}$$

The scheme (4.3.6) satisfies the condition of complete consistency

$$\varrho = \frac{1 - a_2}{1 + a_1}. \tag{4.3.7}$$

From the strong stability of the scheme (4.3.6) for $r \leq 1/2$, it follows that

$$\varrho_0 = \max \left\{ \left| \frac{1 - \tau}{1 + \tau} \right|, \left| \frac{1 - 4r + \tau}{1 + \tau} \right| \right\}, \tag{4.3.8}$$

$$\varrho_1 = 1 - h^2, \tag{4.3.9}$$

$$m \simeq \left|\ln \varepsilon\right| N^2. \tag{4.3.10}$$

Thus, scheme (4.3.6) converges twice as fast as the explicit scheme. The following scheme is quite analogous:

$$\frac{u^{n+1} - u^n}{\tau} = \Lambda_1 u^n + \Lambda_2 u^{n+1}.$$

(iii) *The scheme of upper relaxation* (u.r. scheme). This is due to Young and Frankel [44, 45].

$$(E - \alpha \tau \Omega_1) \frac{u^{n+1} - u^n}{\tau} = \Lambda u^n,$$
$$\Omega_1 = \frac{T_{-1} + T_{-2} - 4E}{h^2}, \qquad \Lambda = \Lambda_1 + \Lambda_2. \tag{4.3.11}$$

The u.r. scheme has complete consistency

$$\varrho = \frac{1 + 2\alpha r + (2\alpha - 4) r \left(\sin^2 \frac{k_1 h}{2} + \sin^2 \frac{k_2 h}{2}\right) + \alpha r (\sin k_1 h + \sin k_2 h)\, i}{1 + 2\alpha r + 2\alpha r \left(\sin^2 \frac{k_1 h}{2} + \sin^2 \frac{k_2 h}{2}\right) + \alpha r (\sin k_1 h + \sin k_2 h)\, i}. \tag{4.3.12}$$

The scheme (4.3.11) is strongly stable for the condition $\alpha > 1$. The expression for ϱ is a complex number because functions $\sin k_1 x_1 \sin k_2 x_2$ are not the eigen-functions of the Ω_1 operator and the harmonic analysis of stability ceases to be rigorous. It is easy to show that $\varrho = 1 - \text{const} \cdot h$ for an appropriate selection of α and τ [53, 27]. It is clear that the u.r. scheme increases the convergence speed by one order as compared with the preceding schemes. The u.r. scheme is realized by a recurrence procedure along the principal diagonal.

(iv) *The scheme of alternating directions* (a.d. scheme)

$$\begin{aligned} \frac{u^{n+1/2} - u^n}{\tau} &= \frac{1}{2} (\Lambda_1 u^{n+1/2} + \Lambda_2 u^n); \\ \frac{u^{n+1} - u^{n+1/2}}{\tau} &= \frac{1}{2} (\Lambda_1 u^{n+1/2} + \Lambda_2 u^{n+1}). \end{aligned} \tag{4.3.13}$$

The equivalent scheme in whole steps is

$$\left.\begin{aligned} \frac{u^{n+1} - u^n}{\tau} &= \Omega_1 u^{n+1} + \Omega_2 u^n, \\ \Omega_1 &= \frac{\Lambda_1 + \Lambda_2}{2} - \frac{\tau}{4} \Lambda_1 \Lambda_2, \\ \Omega_2 &= \frac{\Lambda_1 + \Lambda_2}{2} + \frac{\tau}{4} \Lambda_1 \Lambda_2. \end{aligned}\right\} \tag{4.3.14}$$

From Eq. (4.3.14) follows complete consistency

$$\varrho = \frac{\left(1 - \frac{1}{2} a_1\right)\left(1 - \frac{1}{2} a_2\right)}{\left(1 + \frac{1}{2} a_1\right)\left(1 + \frac{1}{2} a_2\right)}. \tag{4.3.15}$$

The a.d. scheme is strongly stable for any τ

$$\varrho_0 = \max \left\{ \left(\frac{1 - \frac{1}{2}\tau}{1 + \frac{1}{2}\tau} \right)^2, \left(\frac{1 - 2r\left(1 - \frac{h^2}{4}\right)}{1 + 2r\left(1 - \frac{h^2}{4}\right)} \right)^2 \right\}, \tag{4.3.16}$$

$$\varrho_1 = 1 - 2h, \tag{4.3.17}$$

$$m \simeq \frac{1}{2} \left|\ln \varepsilon\right| N. \tag{4.3.18}$$

The a. d. scheme, as with the u.r. scheme, increases the rate of convergence by one order as compared with schemes (i) and (ii). This is related to the strong stability of the a.d. scheme for any τ, as opposed to schemes (i) and (ii), which are stable for sufficiently small τ (of order of h^2).

(v) *The scheme of stabilizing corrections* (s.c. scheme)

$$\frac{u^{n+1/2} - u^n}{\tau} = \Lambda_1 u^{n+1/2} + \Lambda_2 u^n, \qquad \frac{u^{n+1} - u^{n+1/2}}{\tau} = \Lambda_2 (u^{n+1} - u^n). \tag{4.3.19}$$

The equivalent scheme in whole steps is

$$\frac{u^{n+1} - u^n}{\tau} = \Omega_1 u^{n+1} + \Omega_2 u^n, \qquad \Omega_1 = \Lambda_1 + \Lambda_2 - \tau \Lambda_1 \Lambda_2, \quad \Omega_2 = \tau \Lambda_1 \Lambda_2, \tag{4.3.20}$$

$$\varrho = \frac{1 + a_1 a_2}{1 + a_1 + a_2 + a_1 a_2} = \frac{1 + a_1 a_2}{(1 + a_1)(1 + a_2)}, \tag{4.3.21}$$

$$\varrho_0 = \max \left\{ \frac{1 + \tau^2}{1 + 2\tau + \tau^2}, \frac{1 + 16r^2\left(1 - \frac{h^2}{2}\right)}{1 + 8r\left(1 - \frac{h^2}{4}\right) + 16r^2\left(1 - \frac{h^2}{2}\right)} \right\}, \tag{4.3.22}$$

$$\varrho_1 = \frac{1}{1 + h} \simeq 1 - h, \tag{4.3.23}$$

$$m \simeq |\ln \varepsilon| N. \tag{4.3.24}$$

(vi) *The splitting scheme*

$$\frac{u^{n+1/2} - u^n}{\tau} = \Lambda_1 (\alpha u^{n+1/2} + \beta u^n); \qquad \frac{u^{n+1} - u^{n+1/2}}{\tau} = \Lambda_2 (\alpha u^{n+1} + \beta u^{n+1/2}). \tag{4.3.25}$$

The equivalent scheme in whole steps is

$$\left.\begin{aligned} \frac{u^{n+1} - u^n}{\tau} &= \Omega_1 u^{n+1} + \Omega_2 u^n, \\ \Omega_1 &= \alpha (\Lambda_1 + \Lambda_2) - \alpha^2 \tau \Lambda_1 \Lambda_2, \\ \Omega_2 &= \beta (\Lambda_1 + \Lambda_2) + \beta^2 \tau \Lambda_1 \Lambda_2. \end{aligned}\right\} \tag{4.3.26}$$

The scheme (4.3.25) satisfies the condition of complete consistency for $\alpha = \beta = 1/2$. At the same time it becomes equivalent to the a.d. scheme.

(vii) *The predictor-corrector scheme* (scheme of approximation corrections).

$$\frac{u^{n+1/4} - u^n}{\tau/2} = \Lambda_1 u^{n+1/4}; \qquad \frac{u^{n+1/2} - u^{n+1/4}}{\tau/2} = \Lambda_2 u^{n+1/2}; \qquad \frac{u^{n+1} - u^n}{\tau} = (\Lambda_1 + \Lambda_2) u^{n+1/2} = \Lambda u^{n+1/2}. \tag{4.3.27}$$

From the first two equations

$$A\, u^{n+1/2} = \left(E - \frac{\tau}{2} \Lambda_1\right)\left(E - \frac{\tau}{2} \Lambda_2\right) u^{n+1/2} = E\, u^n. \tag{4.3.28}$$

By eliminating $u^{n+1/2}$ between the last of Eqs. (4.3.27) and Eq. (4.3.28) we find that

$$A\,(u^{n+1} - u^n) = \tau\, \Lambda\, u^n. \tag{4.3.29}$$

It is easy to see that the scheme of approximation corrections (a.c. scheme) is equivalent to the a.d. scheme.

(viii) *Schemes with singular operators.* V. K. Saul'ev [27], N. I. Buleev [20], A. A. Samarskii [50], and V. P. Il'in [48, 100] have applied integration schemes and iteration schemes which can be considered as alternative forms of schemes with fractional steps for (4.1.3). In application to the heat conduction equation and to Laplace's equation, in place of the approximation

$$\Lambda = \Lambda_1 + \Lambda_2, \qquad \Lambda_i = \frac{\Delta_i \Delta_{-i}}{h_i^2} = \frac{T_i - E}{h_i^2} - \frac{E - T_i^{-1}}{h_i^2}$$

we use the approximation

$$\left.\begin{aligned} \Lambda &= \Omega_1 + \Omega_2, \\ \Omega_1 &= \frac{T_1^{-1} - E}{h_1^2} + \frac{T_2^{-1} - E}{h_2^2}, \\ \Omega_2 &= \frac{T_1 - E}{h_1^2} + \frac{T_2 - E}{h_2^2}. \end{aligned}\right\} \tag{4.3.30}$$

Since for any sufficiently smooth function f we have

$$\|\Omega_i f\| = O\left(\frac{1}{h}\right); \quad \|\Omega_1 \Omega_2 f\| = O\left(\frac{1}{h_1 h_2}\right), \quad i = 1, 2, \tag{4.3.31}$$

the operators Ω_1, Ω_2, $\Omega_1\,\Omega_2$ are singular (see Sec. 1.2). The integration scheme of Eq. (4.1.3) is based on expression (4.3.30) and in fractional steps is

$$\left.\begin{aligned} \frac{u^{n+1/2} - u^n}{\tau} &= \Omega_1(\alpha\, u^{n+1/2} + \beta\, u^n); \\ \frac{u^{n+1} - u^{n+1/2}}{\tau} &= \Omega_2(\alpha\, u^{n+1} + \beta\, u^{n+1/2}), \\ \alpha \geq 0, \quad \beta &\geq 0, \quad \alpha + \beta = 1, \end{aligned}\right\} \tag{4.3.32}$$

or

$$\begin{aligned} \frac{u^{n+1/2} - u^n}{\tau} &= \alpha\, \Omega_1\, u^{n+1/2} + \beta\, \Omega_2\, u^n; \\ \frac{u^{n+1} - u^{n+1/2}}{\tau} &= \beta\, \Omega_1\, u^{n+1/2} + \alpha\, \Omega_2\, u^{n+1}. \end{aligned} \tag{4.3.33}$$

Eqs. (4.3.32) have the structure of the splitting scheme, and Eqs. (4.3.33) have the structure of the a.d. scheme. Schemes (4.3.32) and (4.3.33) are equivalent because they have identical schemes in whole steps

$$(E - \alpha \tau \Omega_1)(E - \alpha \tau \Omega_2) u^{n+1} = (E + \beta \tau \Omega_1)(E + \beta \tau \Omega_2) u^n. \quad (4.3.34)$$

After reduction of scheme (4.3.34) to

$$\begin{aligned}\frac{u^{n+1} - u^n}{\tau} &= \alpha(\Omega_1 + \Omega_2) u^{n+1} + \beta(\Omega_1 + \Omega_2) u^n + \tau \Omega_1 \Omega_2 (\beta^2 u^n - \alpha^2 u^{n+1}) \\ &= \Lambda(\alpha u^{n+1} + \beta u^n) + \tau \Omega_1 \Omega_2 (\beta^2 u^n - \alpha^2 u^{n+1}) \qquad (4.3.35)\end{aligned}$$

it is easy to estimate the order of approximation of scheme (4.3.34). It follows from Eqs. (4.3.31) that scheme (4.3.35) does not approximate Eq. (4.1.3) for $\beta \neq \alpha$ and when we proceed to the limit with $\tau/h = \text{const}$. On the other hand, consistency is attained for $\beta = \alpha = 1/2$, but it is of the first order instead of the second.

Treating the expression (4.3.35) as an iteration scheme, we see that it satisfies the property of complete consistency for $\alpha = \beta = 1/2$. From the expression for ϱ,

$$\varrho = \frac{1 - \beta(a_1 + a_2) + 4r^2\beta^2\left(\sin^2\frac{k_1 h}{2} + \sin^2\frac{k_2 h}{2} + 2\sin\frac{k_1 h}{2}\sin\frac{k_2 h}{2}\cos\frac{k_2 - k_1}{2}h\right)}{1 + \alpha(a_1 + a_2) + 4r^2\alpha^2\left(\sin^2\frac{k_1 h}{2} + \sin^2\frac{k_2 h}{2} + 2\sin\frac{k_1 h}{2}\sin\frac{k_2 h}{2}\cos\frac{k_2 - k_1}{2}h\right)}, \quad (4.3.36)$$

it follows that the scheme is strongly stable for $\alpha \geq 1/2$.

The execution of Eqs. (4.3.32) and (4.3.33) is very simple: a recurrent calculation from below upwards and from left to right is applied at the first fractional step, and a reverse direction is used at the second fractional step.

Schemes with singular parameters are carried out in a similar way to the execution of symmetrical schemes of upper relaxation (see, for example, [39]).

(ix) *Schemes with additional parameters.* It is evident that additional parameters can be introduced into the stabilizing correction scheme of Douglas and Rachford. This was done by Douglas in [26]. The s.c. scheme with arbitrary parameter is

$$\begin{aligned}\frac{u^{n+1/2} - u^n}{\tau} &= \alpha \Lambda_1 u^{n+1/2} + (\Lambda - \alpha \Lambda_1) u^n; \\ \frac{u^{n+1} - u^{n+1/2}}{\tau} &= \alpha \Lambda_2 (u^{n+1} - u^n).\end{aligned} \qquad (4.3.37)$$

After eliminating $u^{n+1/2}$ we have

$$(E - \alpha\tau\Lambda_1)(E - \alpha\tau\Lambda_2) u^{n+1} = [E + \tau(1-\alpha)\Lambda + \alpha^2\tau^2\Lambda_1\Lambda_2] u^n. \quad (4.3.38)$$

For ϱ the following expression is obtained

$$\varrho = \frac{1 - (1 - \alpha)(a_1 + a_2) + \alpha^2 a_1 a_2}{1 + \alpha(a_1 + a_2) + \alpha^2 a_1 a_2}. \tag{4.3.39}$$

The scheme (4.3.37) has complete consistency and is strongly stable for $\alpha \geq 1/2$. For $\alpha = 1$ scheme (4.3.37) represents the s.c. scheme of Douglas-Rachford, and for $\alpha = 1/2$ the s.c. scheme of Douglas-Rachford is equivalent to the a.d. scheme.

V. P. Il'in [48] has proposed another one-parameter set of schemes which include the a.d. schemes and also the s.c. scheme of Douglas-Rachford:

$$\begin{aligned} \frac{u^{n+1/2} - u^n}{\tau} &= \frac{1}{2}(\Lambda_1 u^{n+1/2} + \Lambda_2 u^n); \\ \frac{u^{n+1} - u^{n+1/2}}{\tau} &= k\frac{u^{n+1/2} - u^n}{\tau} + \frac{1}{2}\Lambda_2(u^{n+1} - u^n). \end{aligned} \tag{4.3.40}$$

The scheme in whole steps is

$$\left(E - \frac{1}{2}\tau\Lambda_1\right)\left(E - \frac{1}{2}\tau\Lambda_2\right)\frac{u^{n+1} - u^n}{\tau} = \frac{1+k}{2}\Lambda u^n. \tag{4.3.41}$$

Scheme (4.3.41) satisfies the property of complete consistency. From the expression

$$\varrho = \frac{1 - \frac{k}{2}(a_1 + a_2) + \frac{1}{4}a_1 a_2}{1 + \frac{1}{2}(a_1 + a_2) + \frac{1}{4}a_1 a_2} \tag{4.3.42}$$

it follows that Eq. (4.3.40) is strongly stable for $-1 < k \leq 1$. For $k = 0$ scheme (4.3.40) is the s.c. scheme of Douglas-Rachford, and for $k = 1$ it is identical to the a.d. scheme.

V. A. Enal'skii [49] analyzed a one-parameter set of schemes containing the splitting and a.d. schemes for particular values of the parameters. This set of schemes is of the type

$$\begin{aligned} \frac{u^{n+1/2} - u^n}{\tau} &= \alpha\Lambda_1 u^{n+1/2} + \beta\Lambda_1 u^n + \gamma\Lambda_2 u^n; \\ \frac{u^{n+1} - u^{n+1/2}}{\tau} &= \alpha\Lambda_2 u^{n+1} + \gamma\Lambda_1 u^{n+1/2} + \beta\Lambda_2 u^{n+1/2}, \end{aligned} \tag{4.3.43}$$

where α, β, γ are undetermined parameters.

After eliminating $u^{n+1/2}$ we get the scheme

$$\left.\begin{aligned} \frac{u^{n+1} - u^n}{\tau} &= \Omega_1 u^{n+1} + \Omega_2 u^n, \\ \Omega_1 &= \alpha\Lambda - \alpha^2\tau\Lambda_1\Lambda_2, \\ \Omega_2 &= (\beta + \gamma)\Lambda + \tau[\beta\gamma\Lambda^2 + (\beta - \gamma)^2\Lambda_1\Lambda_2]. \end{aligned}\right\} \tag{4.3.44}$$

With the conditions

$$\alpha + \beta + \gamma = 1; \quad (\beta - \gamma)^2 - \alpha^2 = 0 \tag{4.3.45}$$

Eq. (4.3.44) is

$$\frac{u^{n+1} - u^n}{\tau} = \Lambda u^n + \alpha \Lambda (u^{n+1} - u^n) + \beta \gamma \tau \Lambda^2 u^n - \alpha^2 \tau \Lambda_1 \Lambda_2 (u^{n+1} - u^n).$$

Thus, scheme (4.3.43) is reduced to the canonical form

$$B \frac{u^{n+1} - u^n}{\tau} = \Lambda u^n, \qquad B = (E + \beta \gamma \tau \Lambda)^{-1} (E - \alpha \tau \Lambda_1)(E - \alpha \tau \Lambda_2) \tag{4.3.46}$$

and satisfies the condition of complete consistency.

From the second of Eqs. (4.3.45) follows the alternative

$$\alpha = \beta - \gamma; \qquad \alpha = \gamma - \beta.$$

Since the parameters α, β, γ are related by the two relations (4.3.45), scheme (4.3.43) allows two arbitrary parameters τ and α. Assuming that $\gamma = 0$, we find that $\alpha = \beta = 1/2$, i.e., a splitting scheme with two equal weights is obtained; putting $\beta = 0$ we have $\alpha = \gamma = 1/2$, i.e., an a.d. scheme is obtained.

It is easy to establish that these schemes can be represented in the form

$$B \frac{u^{n+1} - u^n}{\tau} = \Lambda u^n, \tag{4.3.47}$$

where

$$B = (E - \alpha \tau \Lambda_1)(E - \alpha \tau \Lambda_2) \tag{4.3.48}$$

for a.d., s.c., and a.c. schemes;

$$B = (E - \alpha \tau \Omega_1)(E - \alpha \tau \Omega_2) \tag{4.3.49}$$

for the singular operator scheme; and

$$B = (E + \beta \gamma \tau \Lambda)^{-1} (E - \alpha \tau \Lambda_1)(E - \alpha \tau \Lambda_2) \tag{4.3.50}$$

for the scheme of V. A. Enal'skii. The canonical form (4.3.47) is also included in the works of E. G. D'yakonov [90, 95] and A. A. Samarskii [91].

4.4 Iterative schemes for Laplace's equation in three dimensions

A brief review of the iterative scheme for Laplace's equation in three dimensions is given below. It should be noted here that many properties and classifications of schemes change on going from the equation in two dimensions to the equation in three dimensions.

(i) *The a.d. scheme*

$$\left.\begin{aligned} \frac{u^{n+1/3} - u^n}{\tau} &= \frac{1}{3} (\Lambda_1 u^{n+1/3} + \Lambda_2 u^n + \Lambda_3 u^n); \\ \frac{u^{n+2/3} - u^{n+1/3}}{\tau} &= \frac{1}{3} (\Lambda_1 u^{n+1/3} + \Lambda_2 u^{n+2/3} + \Lambda_3 u^{n+1/3}); \\ \frac{u^{n+1} - u^{n+2/3}}{\tau} &= \frac{1}{3} (\Lambda_1 u^{n+2/3} + \Lambda_2 u^{n+2/3} + \Lambda_3 u^{n+1}) \end{aligned}\right\} \tag{4.4.1}$$

is conditionally stable (see Sec. 2.1). The scheme satisfies, as before, the property of complete consistency. However, this is not in the sense of Eq. (4.2.12), but in that of Eq. (4.2.14). In fact we have, after eliminating $u^{n+1/3}$ and $u^{n+2/3}$,

$$\left.\begin{aligned} &B\frac{u^{n+1}-u^n}{\tau} = \Lambda u^n, \\ &B = \left[E + \frac{1}{9}\tau\Lambda + \frac{1}{27}\tau^2(\Lambda_1\Lambda_2 + \Lambda_1\Lambda_3 + \Lambda_2\Lambda_3)\right]^{-1} \times \\ &\quad \times \left(E - \frac{1}{3}\tau\Lambda_1\right)\left(E - \frac{1}{3}\tau\Lambda_2\right)\left(E - \frac{1}{3}\tau\Lambda_3\right). \end{aligned}\right\} \quad (4.4.2)$$

(ii) *The splitting scheme*

$$\left.\begin{aligned} \frac{u^{n+1/3}-u^n}{\tau} &= \Lambda_1(\alpha\, u^{n+1/3} + \beta\, u^n); \\ \frac{u^{n+2/3}-u^{n+1/3}}{\tau} &= \Lambda_2(\alpha\, u^{n+2/3} + \beta\, u^{n+1/3}); \\ \frac{u^{n+1}-u^{n+2/3}}{\tau} &= \Lambda_3(\alpha\, u^{n+1} + \beta\, u^{n+2/3}) \end{aligned}\right\} \quad (4.4.3)$$

in whole steps is

$$\begin{aligned} \frac{u^{n+1}-u^n}{\tau} = {} & (\Lambda_1 + \Lambda_2 + \Lambda_3)(\alpha\, u^{n+1} + \beta\, u^n) - \\ & - \tau(\Lambda_1\Lambda_2 + \Lambda_1\Lambda_3 + \Lambda_2\Lambda_3)(\alpha^2 u^{n+1} - \beta^2 u^n) + \\ & + \tau^2\Lambda_1\Lambda_2\Lambda_3(\alpha^3 u^{n+1} + \beta^3 u^n). \end{aligned} \quad (4.4.4)$$

It follows from this that the splitting scheme does not satisfy the property of a complete approximation for any α. From the expression for ϱ,

$$\varrho = \frac{1 - \beta(a_1 + a_2 + a_3) + \beta^2(a_1 a_2 + a_1 a_3 + a_2 a_3) - \beta^3 a_1 a_2 a_3}{1 + \alpha(a_1 + a_2 + a_3) + \alpha^2(a_1 a_2 + a_1 a_3 + a_2 a_3) + \alpha^3 a_1 a_2 a_3}, \quad (4.4.5)$$

it follows that the scheme is strongly stable for $\alpha \geq 1/2$.

Thus, while the a.d. scheme conserves its property of complete consistency and loses that of absolute strong stability, the splitting scheme loses the property of complete consistency and preserves that of absolute stability. The schemes cease to be equivalent.

(iii) *The approximation corrections scheme* of Brian [25] (see Sec. 2.7)

$$\left.\begin{aligned} \frac{u^{n+1/6}-u^n}{\tau/2} &= \Lambda_1 u^{n+1/6} + \Lambda_2 u^n + \Lambda_3 u^n; \\ \frac{u^{n+2/6}-u^{n+1/6}}{\tau/2} &= \Lambda_2(u^{n+2/6} - u^n); \\ \frac{u^{n+3/6}-u^{n+2/6}}{\tau/2} &= \Lambda_3(u^{n+3/6} - u^n); \\ \frac{u^{n+1}-u^n}{\tau} &= \Lambda_1 u^{n+1/6} + \Lambda_2 u^{n+2/6} + \Lambda_3 u^{n+3/6}. \end{aligned}\right\} \quad (4.4.6)$$

After eliminating fractional steps, the scheme has the form

$$\left(E - \frac{1}{2}\tau\Lambda_1\right)\left(E - \frac{1}{2}\tau\Lambda_2\right)\left(E - \frac{1}{2}\tau\Lambda_3\right)\frac{u^{n+1} - u^n}{\tau} = \Lambda u^n. \tag{4.4.7}$$

The scheme possesses complete consistency. Absolute stability follows from the expression for ϱ,

$$\varrho = \frac{1 - \frac{1}{2}(a_1 + a_2 + a_3) + \frac{1}{4}(a_1 a_2 + a_1 a_3 + a_2 a_3) + \frac{1}{8} a_1 a_2 a_3}{1 + \frac{1}{2}(a_1 + a_2 + a_3) + \frac{1}{4}(a_1 a_2 + a_1 a_3 + a_2 a_3) + \frac{1}{8} a_1 a_2 a_3}. \tag{4.4.8}$$

The a.c. splitting scheme has a simpler structure (see Sec. 2.7)

$$\begin{aligned} &\frac{u^{n+1/6} - u^n}{\alpha\tau} = \Lambda_1 u^{n+1/6}; \qquad \frac{u^{n+2/6} - u^{n+1/6}}{\alpha\tau} = \Lambda_2 u^{n+2/6};\\ &\frac{u^{n+1/2} - u^{n+2/6}}{\alpha\tau} = \Lambda_3 u^{n+1/2}; \qquad \frac{u^{n+1} - u^n}{\tau} = \Lambda u^{n+1/2}. \end{aligned} \tag{4.4.9}$$

The scheme in whole steps is

$$(E - \alpha\tau\Lambda_1)(E - \alpha\tau\Lambda_2)(E - \alpha\tau\Lambda_3)\frac{u^{n+1} - u^n}{\tau} = \Lambda u^n. \tag{4.4.10}$$

The scheme has complete consistency and strong stability for $1/2 \leq \alpha \leq 1$. The scheme is transformed into Brian's scheme for $\alpha = 1/2$.

(iv) *The scheme of stabilizing corrections*

$$\left.\begin{aligned} \frac{u^{n+1/3} - u^n}{\tau} &= \alpha\Lambda_1 u^{n+1/3} + (1-\alpha)\Lambda_1 u^n + (\Lambda_2 + \Lambda_3) u^n;\\ \frac{u^{n+2/3} - u^{n+1/3}}{\tau} &= \alpha\Lambda_2 (u^{n+2/3} - u^n);\\ \frac{u^{n+1} - u^{n+2/3}}{\tau} &= \alpha\Lambda_3 (u^{n+1} - u^n). \end{aligned}\right\} \tag{4.4.11}$$

For $\alpha = 1$ we have the s.c. scheme of Douglas-Rachford [12]; and for $\alpha = 1/2$ the s.c. scheme of Douglas [26]. The scheme in whole steps is

$$(E - \alpha\tau\Lambda_1)(E - \alpha\tau\Lambda_2)(E - \alpha\tau\Lambda_3)\frac{u^{n+1} - u^n}{\tau} = \Lambda u^n, \tag{4.4.12}$$

from which complete consistency follows. From the expression for ϱ,

$$\varrho = \frac{1 - (1-\alpha)(a_1 + a_2 + a_3) + \alpha^2(a_1 a_2 + a_1 a_3 + a_2 a_3) + \alpha^3 a_1 a_2 a_3}{1 + \alpha(a_1 + a_2 + a_3) + \alpha^2(a_1 a_2 + a_1 a_3 + a_2 a_3) + \alpha^3 a_1 a_2 a_3}, \tag{4.4.13}$$

strong stability of the s.c. scheme follows for $\alpha \geq 1/2$. The s.c. scheme is equivalent to the a.c. splitting scheme.

(v) *The scheme of universal algorithm.* We can start with the canonical representation

$$(E - \alpha\tau\Lambda_1)(E - \alpha\tau\Lambda_2)(E - \alpha\tau\Lambda_3)\frac{u^{n+1} - u^n}{\tau} = \Lambda u^n \tag{4.4.14}$$

of the schemes with fractional steps, which corresponds to the scheme of universal algorithm (see [27])[1]. It follows from Eq. (4.4.14) that the scheme of universal algorithm is equivalent to the s.c. and a.c. schemes. The scheme also could have the following representation

$$\left.\begin{aligned}&(E - \alpha\tau\Lambda_1)\,u^{n+1/3}\\&\quad = [(E-\alpha\tau\Lambda_1)(E-\alpha\tau\Lambda_2)(E-\alpha\tau\Lambda_3) + \tau\Lambda]u^n;\\&(E-\alpha\tau\Lambda_2)\,u^{n+2/3} = u^{n+1/3};\\&(E-\alpha\tau\Lambda_3)\,u^{n+1} = u^{n+2/3},\end{aligned}\right\} \tag{4.4.15}$$

which corresponds to the scheme of approximate factorization with a corresponding algorithm for the solution of the boundary conditions (see Sec. 2.5 and 2.6).

(vi) *The iterative scheme with additional parameters*, which was suggested by V. P. Il'in, is

$$\left.\begin{aligned}\frac{u^{n+1/3}-u^n}{\tau} &= \Lambda_1 u^{n+1/3} + \Lambda_2 u^n + \Lambda_3 u^n;\\ \frac{u^{n+2/3}-u^{n+1/3}}{\tau} &= k_1\frac{u^{n+1/3}-u^n}{\tau} + \Lambda_2(u^{n+2/3}-u^n);\\ \frac{u^{n+1}-u^{n+2/3}}{\tau} &= k_2\frac{u^{n+2/3}-u^n}{\tau} + \Lambda_3(u^{n+1}-u^n).\end{aligned}\right\} \tag{4.4.16}$$

After eliminating fractional steps we have

$$(E-\tau\Lambda_1)(E-\tau\Lambda_2)(E-\tau\Lambda_3)\frac{u^{n+1}-u^n}{\tau} = (1+k_1)(1+k_2)\Lambda u^n. \tag{4.4.17}$$

Assuming that

$$(1+k_1)(1+k_2) = \frac{1}{\alpha},$$

we see that scheme (4.4.16) is equivalent to the a.c. and s.c. schemes and to the scheme with a stabilizing operator.

(vii) *The scheme with singular operators.* As opposed to the preceding schemes, the scheme with singular operators requires, as before, only two fractional steps.

In the three-dimensional case we have

$$\left.\begin{aligned}\Lambda &= \sum_{i=1}^{3}\Lambda_i = \Omega_1 + \Omega_2;\\ \Omega_1 &= \frac{T_{-1}-E}{h_1^2} + \frac{T_{-2}-E}{h_2^2} + \frac{T_{-3}-E}{h_3^2};\\ \Omega_2 &= \frac{T_1-E}{h_1^2} + \frac{T_2-E}{h_2^2} + \frac{T_3-E}{h_3^2}.\end{aligned}\right\} \tag{4.4.18}$$

[1] Eq. (4.4.14) was shown for the first time in the works of E. G. D'yakonov [95, 90] and A. A. Samarskii [91].

The following equivalent scheme corresponds to representation (4.4.18)

$$\frac{u^{n+1/2} - u^n}{\tau} = \Omega_1(\alpha u^{n+1/2} + \beta u^n),$$
$$\frac{u^{n+1} - u^{n+1/2}}{\tau} = \Omega_2(\alpha u^{n+1} + \beta u^{n+1/2}), \tag{4.4.19}$$

$$\frac{u^{n+1/2} - u^n}{\tau} = \alpha \Omega_1 u^{n+1/2} + \beta \Omega_2 u^n,$$
$$\frac{u^{n+1} - u^{n+1/2}}{\tau} = \beta \Omega_1 u^{n+1/2} + \alpha \Omega_2 u^{n+1}. \tag{4.4.20}$$

4.5 Iterative schemes for elliptic equations

For the elliptic equation

$$L u + f = \sum_{i,j=1}^{m} a_{ij} \frac{\partial^2 u}{\partial x_i \partial x_j} + f = 0 \tag{4.5.1}$$

the parallel between the iterative schemes and integration schemes of the corresponding parabolic equation

$$\frac{\partial u}{\partial t} = \sum_{i,j=1}^{m} a_{ij} \frac{\partial^2 u}{\partial x_i \partial x_j} + f \tag{4.5.2}$$

is always valid. However, a number of schemes possessing the properties of complete consistency and strong stability for Laplace's equation lose these properties in the case of Eq. (4.5.1). A brief analysis of iterative schemes for Eq. (4.5.1) for $f = 0$ is given below.

(i) *The a.d. scheme* is unacceptable because it loses the property of strong stability for $m \geq 2$ (see Sec. 2.4).

(ii) *The splitting scheme* for $m = 2$ (see Sec. 2.4)

$$\frac{u^{n+1/2} - u^n}{\tau} = \Lambda_{11} u^{n+1/2} + \Lambda_{12} u^n;$$
$$\frac{u^{n+1} - u^{n+1/2}}{\tau} = \Lambda_{21} u^{n+1/2} + \Lambda_{22} u^{n+1}$$

is equivalent to the scheme in whole steps

$$\frac{u^{n+1} - u^n}{\tau} = (\Lambda_{11} + \Lambda_{22}) u^{n+1} + 2\Lambda_{12} u^n - \tau(\Lambda_{11} \Lambda_{22} u^{n+1} - \Lambda_{12}^2 u^n)$$

and is strongly stable, but it loses the property of complete consistency. The same holds true for the splitting scheme (2.4.12) at $m = 3$.

(iii) *The scheme of stabilizing corrections*

$$\left.\begin{aligned}
\frac{u^{n+1/m} - u^n}{\tau} &= \Lambda_{11} u^{n+1/m} + (\Omega - \Lambda_{11}) u^n; \\
\frac{u^{n+2/m} - u^{n+1/m}}{\tau} &= \Lambda_{22}(u^{n+2/m} - u^n); \\
&\cdots\cdots\cdots \\
\frac{u^{n+1} - u^{n+(m-1)/m}}{\tau} &= \Lambda_{mm}(u^{n+1} - u^n), \qquad \Omega = \sum_{i,j=1}^{m} \Lambda_{ij},
\end{aligned}\right\} \tag{4.5.3}$$

is equivalent to the scheme in whole steps

$$\left.\begin{aligned}\frac{u^{n+1}-u^n}{\tau} &= \Lambda u^{n+1} + (\Omega - \Lambda) u^n - \tau \sum_{i<j} \Lambda_{ii} \Lambda_{jj} (u^{n+1} - u^n) + \\ &+ \tau^2 \sum_{i<j<k} \Lambda_{ii} \Lambda_{jj} \Lambda_{kk} (u^{n+1} - u^n) + \cdots \\ &+ (-1)^{m-1} \Lambda_{11} \ldots \Lambda_{mm} \tau^{m-1} (u^{n+1} - u^n), \\ \Lambda &= \sum_{i=1}^{m} \Lambda_{ii}, \quad i, j, k = 1, \ldots, m. \end{aligned}\right\} \tag{4.5.4}$$

From this follows complete consistency at any m. For ϱ we have the expression

$$\begin{aligned}\varrho &= \frac{1 - 2\sum_{i<j} l_{ij} + \sum_{i<j} l_{ii} l_{jj} + \sum_{i<j<k} l_{ii} l_{jj} l_{kk} + \cdots + l_{11} \ldots l_{mm}}{(1 + l_{11})(1 + l_{22}) \cdots (1 + l_{mm})} \\ &= \frac{1 - 2\sum_{i<j} l_{ij} + \sum_{i<j} l_{ii} l_{jj} + \cdots + l_{11} \ldots l_{mm}}{1 + \sum_{i=1}^{m} l_{ii} + \sum_{i<j} l_{ii} l_{jj} + \cdots + l_{11} \ldots l_{mm}}, \end{aligned} \tag{4.5.5}$$

where

$$\begin{aligned} l_{ii} &= 4\tau a_{ii} \frac{\sin^2 \frac{k_i h_i}{2}}{h_i^2}; \\ l_{ij} &= 4 \frac{\tau a_{ij}}{h_i h_j} \cos \frac{k_i h_i}{2} \cos \frac{k_j h_j}{2} \sin \frac{k_i h_i}{2} \sin \frac{k_j h_j}{2} = \frac{\tau a_{ij}}{h_i h_j} \sin k_i h_i \sin k_j h_j . \end{aligned} \tag{4.5.6}$$

For $m = 2$ the scheme possesses strong stability if the following elliptic conditions are satisfied

$$a_{11} > 0, \quad a_{22} > 0, \quad a_{11} a_{22} - a_{12}^2 > 0.$$

With stricter limitations the scheme also will be strongly stable for $m \geq 3$ [see Eq. (2.4.13)].

(iv) *The scheme of approximation corrections* (a.c.)

$$\left.\begin{aligned} \frac{u^{n+\frac{1}{2m}} - u^n}{\alpha\tau} = \Lambda_{11} u^{n+\frac{1}{2m}}, \ldots, \frac{u^{n+\frac{m}{2m}} - u^{n+\frac{m-1}{2m}}}{\alpha\tau} = \Lambda_{mm} u^{n+\frac{m}{2m}}, \\ \frac{u^{n+1} - u^n}{\tau} = \Omega u^{n+\frac{m}{2m}}. \end{aligned}\right. \tag{4.5.7}$$

After eliminating fractional steps we get

$$(E - \alpha\tau\Lambda_{11})(E - \alpha\tau\Lambda_{22}) \ldots (E - \alpha\tau\Lambda_{mm}) \frac{u^{n+1} - u^n}{\tau} = \Omega u^n. \tag{4.5.8}$$

It follows from this that the a.c. scheme has the property of complete consistency. From Eq. (4.5.8) the expression for ϱ is obtained,

$$\varrho = \frac{1 - \sum_{i,j=1}^{m} l_{ij} + \alpha \sum_{i=1}^{m} l_{ii} + \alpha^2 \sum_{i<j} l_{ii} l_{jj} + \cdots + \alpha^m l_{11} \ldots l_{mm}}{1 + \alpha \sum_{i=1}^{m} l_{ii} + \alpha^2 \sum_{i<j} l_{ii} l_{jj} + \cdots + \alpha^m l_{11} \ldots l_{mm}}. \tag{4.5.9}$$

In the case $m = 2$ the a.c. scheme is strongly stable for $1/2 \leq \alpha \leq 1$.

(v) *The scheme of majorant operator* (m.o.) [48, 49] for $m = 2$ is

$$(\Lambda_{11} + \Lambda_{22}) \frac{u^{n+1} - u^n}{\tau} = \Omega\, u^n. \tag{4.5.10}$$

The equation

$$(\Lambda_{11} + \Lambda_{22})\, u^{n+1} = [(\Lambda_{11} + \Lambda_{22}) + \tau\, \Omega]\, u^n \tag{4.5.10'}$$

is solved by means of additional (internal) iterations when the right-hand side is given.

Thus, the elliptic difference operator Ω of general form in the m.o. scheme is replaced by a difference analogous to the Laplace operator.

(vi) *The scheme of stabilizing operator* for $m = 2$ is

$$(E - \alpha\, \tau\, \Lambda_{11})\, (E - \alpha\, \tau\, \Lambda_{22}) \frac{u^{n+1} - u^n}{\tau} = \Omega\, u^n. \tag{4.5.11}$$

The scheme (4.5.11) is equivalent to the a.c. scheme and can be executed in the form of the approximate factorization scheme

$$\begin{aligned} (E - \alpha\, \tau\, \Lambda_{11})\, u^{n+1/2} &= [(E - \alpha\, \tau\, \Lambda_{11})\, (E - \alpha\, \tau\, \Lambda_{22}) + \tau\, \Omega]\, u^n; \\ (E - \alpha\, \tau\, \Lambda_{22})\, u^{n+1} &= u^{n+1/2}. \end{aligned} \tag{4.5.12}$$

Notice that schemes of majorant and stabilizing operators are derived from the scheme of universal algorithm

$$B \frac{u^{n+1} - u^n}{\tau} = \Omega\, u^n.$$

In the first case the operator Ω is replaced by the operator B of much simpler structure, and in the second case the operator B is factorized and it satisfies the condition of strong stability at the same time.

For a general scheme using the stabilizing operator see Sec. 9.6.

(vii) *A scheme with diagonal sweeps.* We note here that in the scheme of I. D. Sofronov [65] the predictor-corrector and also diagonal iterations are applicable on the basis of the splitting method.

For simplicity consider the integration region as being a square ($h_1 = h_2 = h$; $N_1 = N_2 = N$), $m = 2$. Assume that

$$L = L_{11} + L_{22} + M_{11} + M_{22}, \tag{4.5.13}$$

where

$$\left.\begin{aligned} L_{11} &= (a_{11} - |a_{12}|) \frac{\partial^2}{\partial x_1^2}; \\ L_{22} &= (a_{22} - |a_{12}|) \frac{\partial^2}{\partial x_2^2}; \\ M_{11} &= |a_{12}| (1 + \sigma) \frac{\partial^2}{\partial \xi_1^2}; \\ M_{22} &= |a_{12}| (1 - \sigma) \frac{\partial^2}{\partial \xi_2^2}; \\ \sigma = \operatorname{sign} a_{12}; \quad \xi_1 &= \frac{1}{\sqrt{2}} (x_1 + x_2); \quad \xi_2 = \frac{1}{\sqrt{2}} (x_1 - x_2). \end{aligned}\right\} \tag{4.5.14}$$

Let Λ_1, Λ_2, Ω_1, Ω_2 be central three-point approximations of operators L_{11}, L_{22}, M_{11}, M_{22} which act along the coordinate and diagonal lines, respectively. The difference splitting scheme, which corresponds to Eq. (4.5.13), is

$$\begin{aligned} \frac{u^{n+1/8} - u^n}{\tau/2} &= \Lambda_1 u^{n+1/8}; & \frac{u^{n+2/8} - u^{n+1/8}}{\tau/2} &= \Lambda_2 u^{n+2/8}; \\ \frac{u^{n+3/8} - u^{n+2/8}}{\tau/2} &= \Omega_1 u^{n+3/8}; & \frac{u^{n+1/2} - u^{n+3/8}}{\tau/2} &= \Omega_2 u^{n+1/2}. \end{aligned} \tag{4.5.15}$$

When

$$a_{11} \geq |a_{12}|; \quad a_{22} \geq |a_{12}| \tag{4.5.16}$$

operators Λ_1, Λ_2 have non-positive eigenvalues and scheme (4.5.15) is strongly stable and possesses accuracy of the first order.

Using the corrector

$$\frac{u^{n+1} - u^n}{\tau} = \Omega u^{n+1/2} = (\Lambda_1 + \Lambda_2 + \Omega_1 + \Omega_2) u^{n+1/2} \tag{4.5.17}$$

we find u^{n+1} with an accuracy of $O(\tau^2 + h^2)$.

I. D. Sofronov in his earlier work [66] considered the splitting scheme with corrector.

4.6 Schemes with variable steps

Up to now we have considered only the iterative schemes with constant step. The norm of the step operator for explicit schemes was of type $(1 - C h^2)$ and for implicit schemes $(1 - C h)$, provided the criteria of strong stability were fulfilled and τ was chosen properly. In both cases the convergence can be speeded up by a selection of the variable iterative parameter τ_n.

L. D. Richardson [52] has obtained a significant improvement in convergence of the explicit scheme (4.3.2) by introducing the variable step. The work of L. D. Richardson as well as other works [53] made it possible to reduce the search for optimal steps to the construction of a polynomial with the lowest deviation from zero in the case of the explicit iterative scheme. Let there be m iterations with variable steps $\tau_1, \tau_2, \ldots, \tau_m$. Then the amplitude of the harmonics $\sin k_1 x_1 \sin k_2 x_2$ is multiplied by

$$P_m(\alpha_1, \ldots, \alpha_m, \mu) = \varrho_1 \varrho_2 \ldots \varrho_m = (1 - \alpha_1 \mu) \ldots (1 - \alpha_m \mu). \quad (4.6.1)$$

Here

$$\alpha_s = 2a^2 \tau_s; \qquad \mu = \frac{2\sin^2 \frac{k_1 h_1}{2}}{h_1^2} + \frac{2\sin^2 \frac{k_2 h_2}{2}}{h_2^2} = \mu(k_1, k_2). \quad (4.6.2)$$

Parameters $\alpha_1, \ldots, \alpha_m$ are selected in such a way that the quantity

$$P(\alpha_1, \ldots, \alpha_m) = \max_{k_1, k_2} \left| P_m[\alpha, \mu(k_1, k_2)] \right|, \qquad \mu_0 \le \mu \le \mu_1,$$
$$\mu_0 = \mu(1, 1), \qquad \mu_1 = \mu(N_1, N_2), \quad (4.6.3)$$

attains a minimum value.

This problem is replaced by another in which the discrete parameter $\mu(k_1, k_2)$ is considered as being a continuous parameter μ which changes within $[\mu(1, 1), \mu(N_1, N_2)]$. It is evident that the selection of the optimal step of iteration is reduced to a selection of the polynomial $P(\alpha, \mu)$ from the family (4.6.1) with the lowest deviation from zero. The following function is the solution of this problem in the interval $[\mu_0, \mu_1]$ (see [54]),

$$P_m(x) = \frac{T_m\left(\frac{\mu_1 + \mu_0 - 2x}{\mu_1 - \mu_0}\right)}{T_m\left(\frac{\mu_1 + \mu_0}{\mu_1 - \mu_0}\right)}, \quad (4.6.4)$$

where $T_m(x) = \cos(m \arccos x)$ are Chebyshev polynomials. It is obvious that for such a selection of steps the convergence is accelerated up to the order of (see [53, 27])

$$1 - O(h). \quad (4.6.5)$$

In the case of implicit schemes the expression for ϱ is more complex. Thus for the a.d. scheme we have

$$\varrho = \varrho_1 \ldots \varrho_m = R(\alpha, \mu, \nu)$$
$$= \frac{1 - \alpha_1 \mu}{1 + \alpha_1 \mu} \cdot \frac{1 - \alpha_1 \nu}{1 + \alpha_1 \nu} \cdot \frac{1 - \alpha_2 \mu}{1 + \alpha_2 \mu} \cdot \frac{1 - \alpha_2 \nu}{1 + \alpha_2 \nu} \cdots \frac{1 - \alpha_m \mu}{1 + \alpha_m \mu} \cdot \frac{1 - \alpha_m \nu}{1 + \alpha_m \nu}$$
$$= R(\alpha, \mu) R(\alpha, \nu), \quad (4.6.6)$$

$$\alpha_s = \frac{1}{2} a^2 \tau_s; \quad \mu = \mu(k_1) = \frac{4}{h_1^2} \sin^2 \frac{k_1 h_1}{2}; \quad \nu = \nu(k_2) = \frac{4}{h_2^2} \sin^2 \frac{k_2 h_2}{2};$$
$$R(\alpha, \mu) = \prod_{s=1}^{m} \frac{1 - \alpha_s \mu}{1 + \alpha_s \mu}; \quad R(\alpha, \nu) = \prod_{s=1}^{m} \frac{1 - \alpha_s \nu}{1 + \alpha_s \nu}. \quad (4.6.7)$$

The optimal selection of steps α_s corresponds to the solution of the variational problem

$$\min_{\alpha} \max_{k_1, k_2} |R(\alpha, \mu, \nu)|, \quad k_1 = 1, \ldots, N_1, \quad k_2 = 1, \ldots, N_2, \quad (4.6.8)$$

which is reduced to the following problem after a transition from discrete parameters $\mu(k_1)$ and $\nu(k_2)$ to continuous parameters μ and ν:

$$\min_{\alpha} \max_{\mu} |R(\alpha, \mu)|; \quad \mu(1) \leq \mu \leq \mu(N_1), \quad (4.6.9a)$$

$$\min_{\alpha} \max_{\nu} |R(\alpha, \nu)|; \quad \nu(1) \leq \nu \leq \nu(N_2). \quad (4.6.9b)$$

The problem (4.6.9) is not yet completely solved but there are useful studies on this subject [55]. Following [10] as well as [27] let us indicate the method of selection of parameters $\alpha_1, \ldots, \alpha_m$ which secures a decrease of the deviation norm by $1/q$ times ($q < 1$) after m steps and let us evaluate m as a function of h. Assume for simplicity

$$h_1 = h_2 = h; \quad 1 \leq k_s \leq N; \quad s = 1, 2; \quad (N + 1)\, h = \pi.$$

Giving $m, h, q < 1$, divide the interval $(1, N)$ into sub-intervals (k_i, k_{i+1}), $k_1 = 1, \ldots, k_m = N$ in such a way that the following conditions are fulfilled

$$q = \frac{1 - \alpha_1 \mu_1}{1 + \alpha_1 \mu_1} = -\frac{1 - \alpha_1 \mu_2}{1 + \alpha_1 \mu_2} = \frac{1 - \alpha_2 \mu_2}{1 + \alpha_2 \mu_2} = \cdots$$

$$\cdots = -\frac{1 - \alpha_{m-1} \mu_m}{1 + \alpha_{m-1} \mu_m} = \frac{1 - \alpha_m \mu_m}{1 + \alpha_m \mu_m}, \quad \mu_s = \mu(k_s). \quad (4.6.10)$$

This is possible because the value

$$q(\alpha, k) = \frac{1 - \alpha\, \mu(k)}{1 + \alpha\, \mu(k)} \quad (4.6.11)$$

for $\alpha > 0$, $k > 0$ does not exceed 1 in modulus and is a monotonically decreasing function of k for fixed α and a similar function of α for fixed k.

The quantity $q(\alpha_s, k)$ never exceeds q in modulus for k in the interval $k_s \leq k \leq k_{s+1}$, and it does not exceed 1 outside of the interval in modulus. Thus, each harmonic decays in amplitude by not less than $1/q$ times after m steps. Let us calculate the value of m. By assuming that h is sufficiently small

$$\mu_s \simeq k_s^2. \quad (4.6.12)$$

From Eq. (4.6.10) it follows that

$$(1 - q) - \alpha_s (1 + q)\, \mu_s = 0;$$
$$(1 + q) - \alpha_s (1 - q)\, \mu_{s+1} = 0. \quad (4.6.13)$$

By eliminating α_s from Eq. (4.6.13) we have

$$\frac{\mu_{s+1}}{\mu_s} = \left(\frac{1 + q}{1 - q}\right)^2. \quad (4.6.14)$$

Using Eq. (4.6.12)

$$\frac{k_{s+1}}{k_s} \approx \frac{1+q}{1-q}, \qquad s = 1, 2, \ldots, m. \tag{4.6.15}$$

Multiplication of Eq. (4.6.15) results in

$$N \simeq \left(\frac{1+q}{1-q}\right)^m, \qquad m \simeq \frac{\ln N}{\ln \dfrac{1+q}{1-q}} \simeq \frac{\ln \dfrac{1}{h}}{\ln \dfrac{1+q}{1-q}} \simeq \frac{\ln \dfrac{1}{h}}{2q}. \tag{4.6.16}$$

Eq. (4.6.16) has the character of an asymptotic evaluation with undetermined constant.

Let us show (see [10]) that in the case of the a.d. scheme with an appropriate selection of steps $\tau_1, \tau_2, \ldots, \tau_m$ the iterative solution of u^n is identical to an accurate solution of the Dirichlet difference problem by means of $m = N$ iterations (square) and $m = N_1 + N_2$ iterations (rectangular). If we select $\tau_1, \tau_2, \ldots, \tau_m$, $m = N_1 + N_2$ from the condition

$$\alpha_1 = \frac{1}{\mu_1}, \qquad \alpha_2 = \frac{1}{\mu_2}, \qquad \ldots, \quad \alpha_{N_1} = \frac{1}{\mu_{N_1}}, \tag{4.6.17a}$$

$$\alpha_{N_1+1} = \frac{1}{\nu_1}, \qquad \alpha_{N_1+2} = \frac{1}{\nu_2}, \qquad \ldots, \quad \alpha_{N_1+N_2} = \frac{1}{\nu_{N_2}}, \tag{4.6.17b}$$

where

$$\mu_{k_1} = \frac{4 \sin^2 \dfrac{k_1 h_1}{2}}{h_1^2}; \qquad \nu_{k_2} = \frac{4 \sin^2 \dfrac{k_2 h_2}{2}}{h_2^2},$$

then

$$R(\alpha, \mu, \nu) = \varrho_1 \varrho_2 \ldots \varrho_m = 0 \tag{4.6.18}$$

for any k_1 and k_2.

In the case of a square when $h_1 = h_2 = h$, $N_1 = N_2 = N$ it is sufficient to satisfy conditions (4.6.17a), i.e., only N iterations are needed.

4.7 Iterative schemes based on integration schemes for hyperbolic equations

Let us set a correspondence between the equation of damped oscillations and Laplace's equation $\Delta \varphi = 0$

$$b \frac{\partial \Phi}{\partial t} + \frac{\partial^2 \Phi}{\partial t^2} = a^2 \Delta \Phi, \qquad b > 0. \tag{4.7.1}$$

Consider now a system replacing Eq. (4.7.1), by assuming that

$$\frac{\partial \Phi}{\partial x_1} = u_1; \qquad \frac{\partial \Phi}{\partial x_2} = u_2; \qquad \frac{\partial \Phi}{\partial t} = -q = a^2 v. \tag{4.7.2}$$

From Eq. (4.7.2) we have

$$\frac{\partial u_1}{\partial t} + \frac{\partial q}{\partial x_1} = 0; \quad \frac{\partial u_2}{\partial t} + \frac{\partial q}{\partial x_2} = 0; \quad \frac{\partial q}{\partial t} = -a^2\left(\frac{\partial u_1}{\partial x_1} + \frac{\partial u_2}{\partial x_2}\right) - bq. \tag{4.7.3}$$

In the case of v we obtain

$$\frac{\partial u_1}{\partial t} - a^2 \frac{\partial v}{\partial x_1} = 0; \quad \frac{\partial u_2}{\partial t} - a^2 \frac{\partial v}{\partial x_2} = 0; \quad \frac{\partial v}{\partial t} = \frac{\partial u_1}{\partial x_1} + \frac{\partial u_2}{\partial x_2} - b v. \tag{4.7.4}$$

The following system is integrated at the first fractional step

$$\frac{1}{2}\frac{\partial u_1}{\partial t} - a^2 \frac{\partial v}{\partial x_1} = 0; \quad \frac{1}{2}\frac{\partial u_2}{\partial t} = 0; \quad \frac{1}{2}\frac{\partial v}{\partial t} = \frac{\partial u_1}{\partial x_1} \tag{4.7.5}$$

with the use of a majorant implicit scheme of running calculation, which in terms of variables u_1, u_2 and v is

$$K_1 f^{n+1/2} = M_1 f^n, \tag{4.7.6}$$

where

$$K_1 = \begin{Vmatrix} 2\Phi_1 \Psi_1 & 0 & 0 \\ 0 & E & 0 \\ 0 & 0 & 2\Phi_1 \Psi_1 \end{Vmatrix}; \quad M_1 = \begin{Vmatrix} \Phi_1 + \Psi_1 & 0 & a(\Phi_1 - \Psi_1) \\ 0 & E & 0 \\ \frac{1}{a}(\Phi_1 - \Psi_1) & 0 & \Phi_1 + \Psi_1 \end{Vmatrix};$$

$$\Phi_1 = E + \frac{a\tau}{h_1}\Delta_{-1}; \quad \Psi_1 = E - \frac{a\tau}{h_1}\Delta_1; \quad f = (u_1, u_2, v). \tag{4.7.7}$$

At the second fractional step the system

$$\frac{1}{2}\frac{\partial u_1}{\partial t} = 0; \quad \frac{1}{2}\frac{\partial u_2}{\partial t} - a^2 \frac{\partial v}{\partial x_2} = 0; \quad \frac{1}{2}\frac{\partial v}{\partial t} = \frac{\partial u_2}{\partial x_2} \tag{4.7.8}$$

is integrated with the use of an analogous scheme

$$K_2 f^{n+1} = M_2 f^{n+1/2}, \tag{4.7.9}$$

where

$$K_2 = \begin{Vmatrix} E & 0 & 0 \\ 0 & 2\Phi_2 \Psi_2 & 0 \\ 0 & 0 & 2\Phi_2 \Psi_2 \end{Vmatrix}; \quad M_2 = \begin{Vmatrix} E & 0 & 0 \\ 0 & \Phi_2 + \Psi_2 & a(\Phi_2 - \Psi_2) \\ 0 & \frac{1}{a}(\Phi_2 - \Psi_2) & \Phi_2 + \Psi_2 \end{Vmatrix}; \tag{4.7.10}$$

$$\Phi_2 = E + \frac{a\tau}{h_2}\Delta_{-2}; \quad \Psi_2 = E - \frac{a\tau}{h_2}\Delta_2; \quad f = (u_1, u_2, v).$$

It is easy to establish that formulas (4.7.6) and (4.7.9) are equivalent to formulas (3.4.2) and (3.4.3).

After this the corrector is applied

$$\frac{f^{n+2} - f^n}{2\tau} = \Omega f^n, \tag{4.7.11}$$

where

$$\Omega = \left\| \begin{matrix} 0 & 0 & -a^2 \dfrac{\Delta_1 + \Delta_{-1}}{2h_1} \\ 0 & 0 & -a^2 \dfrac{\Delta_2 + \Delta_{-2}}{2h_2} \\ \dfrac{\Delta_1 + \Delta_{-1}}{2h_1} & \dfrac{\Delta_2 + \Delta_{-2}}{2h_2} & -b \end{matrix} \right\| . \tag{4.7.12}$$

The scheme of stabilizing operator can also be used:

$$K_1 K_2 \frac{f^{n+2} - f^n}{2\tau} = \Omega f^n. \tag{4.7.13}$$

4.8 Solution of the boundary value problem for Poisson's equation

It is shown below that the solution of the boundary value problem for Poisson's equation even with the use of strongly stable and completely consistent systems requires a special approximation of the right-hand side. (This was pointed out by E. G. D'yakonov and A. A. Samarskii.)

For Poisson's equation in region G: $\{0 < x_i < \pi,\ i = 1, 2\}$

$$\Delta u + q = 0; \quad \Delta = \frac{\partial^2}{\partial x_1^2} + \frac{\partial^2}{\partial x_2^2}, \tag{4.8.1}$$

with boundary conditions of the first kind

$$u(s) = f(s); \quad (x_1(s), x_2(s)) \in \gamma \tag{4.8.2}$$

the weighted splitting scheme can be applied

$$\begin{aligned} \frac{u^{n+1/2} - u^n}{\tau} &= \Lambda_1 (\alpha u^{n+1/2} + \beta u^n) + q_1; \\ \frac{u^{n+1} - u^{n+1/2}}{\tau} &= \Lambda_2 (\alpha u^{n+1} + \beta u^{n+1/2}) + q_2, \end{aligned} \tag{4.8.3}$$

where q_1, q_2 are some right-hand terms which are still undetermined.

The scheme in whole steps is

$$(E - \alpha\tau\Lambda_1)(E - \alpha\tau\Lambda_2) u^{n+1} = (E + \beta\tau\Lambda_1)(E + \beta\tau\Lambda_2) u^n + \tau Q, \tag{4.8.4}$$

where

$$Q = B_2 q_1 + A_1 q_2 = (E + \beta\tau\Lambda_2) q_1 + (E - \alpha\tau\Lambda_1) q_2 .$$

In the case of Laplace's equation it is necessary that

$$Q = q. \tag{4.8.5}$$

Assuming that $q_1 = 0$ we have

$$(E - \alpha\tau\Lambda_1) q_2 = q. \tag{4.8.6}$$

Since $\|E - \alpha\tau\Lambda_1\|^{-1} \le 1$, Eq. (4.8.6) is soluble by ordinary sweep.

4.9 Iterative schemes with averaging

Only convergence of iterations for whole steps has been considered so far. Let us show now that in some cases it is of advantage to consider quantities in whole and fractional steps simultaneously. To the boundary value problem of elliptic type

$$L u + q = 0; \qquad u(s) = f(s) \tag{4.9.1}$$

a corresponding iterative scheme is indicated

$$\frac{u^{n+1/2} - u^n}{\tau} = \Lambda_1(\alpha\, u^{n+1/2} + \beta\, u^n) + q_1, \qquad \alpha + \beta = 1, \tag{4.9.2a}$$

$$\frac{u^{n+1} - u^{n+1/2}}{\tau} = \Lambda_2(\alpha\, u^{n+1} + \beta\, u^{n+1/2}) + q_2, \tag{4.9.2b}$$

$$\Lambda = \Lambda_1 + \Lambda_2 \sim L. \tag{4.9.3}$$

Operators Λ_1, Λ_2 do not contain the parameter τ, and consistency of Eq. (4.9.3) is complete. Now extend Eqs. (4.9.2)

$$\frac{u^{n+3/2} - u^{n+1}}{\tau} = \Lambda_1(\alpha\, u^{n+3/2} + \beta\, u^{n+1}) + q_1. \tag{4.9.4}$$

By adding Eq. (4.9.2a) to Eq. (4.9.2b) and Eq. (4.9.2b) to Eq. (4.9.4) we obtain

$$\frac{u^{n+1} - u^n}{\tau} = \Lambda_1(\alpha\, v^n + \beta\, u^n) + \Lambda_2(\alpha\, u^{n+1} + \beta\, v^n) + Q, \tag{4.9.5a}$$

$$\frac{v^{n+1} - v^n}{\tau} = \Lambda_1(\alpha\, v^{n+1} + \beta\, u^{n+1}) + \Lambda_2(\alpha\, u^{n+1} + \beta\, v^n) + Q, \tag{4.9.5b}$$

where

$$v^n = u^{n+1/2}; \qquad v^{n+1} = u^{n+3/2}; \qquad Q = q_1 + q_2. \tag{4.9.6}$$

Finally, adding Eqs. (4.9.5a) and (4.9.5b)

$$\begin{aligned}\frac{(u^{n+1} + v^{n+1}) - (u^n + v^n)}{\tau} = {} & \Lambda_1[\alpha(v^n + v^{n+1}) + \beta(u^n + u^{n+1})] + \\ & + 2\Lambda_2(\alpha\, u^{n+1} + \beta\, v^n) + 2Q.\end{aligned} \tag{4.9.7}$$

On the assumption that scheme (4.9.2) is strongly stable

$$u^n \to u; \qquad v^n \to v, \tag{4.9.8}$$

and from Eqs. (4.9.7) and (4.9.8) we have

$$\Lambda_1(\alpha\, v + \beta\, u) + \Lambda_2(\alpha\, u + \beta\, v) + Q = 0.$$

For $\alpha = 1/2$

$$\Lambda\left(\frac{u + v}{2}\right) + Q = 0. \tag{4.9.9}$$

If we assume that

$$Q = q_1 + q_2 = q, \tag{4.9.10}$$

then Eq. (4.9.9) means that the average value $(u^n + u^{n+1/2})/2$ is reduced to an exact solution of the difference boundary value problem

$$\Lambda u + q = 0; \qquad u(s) = f(s). \tag{4.9.11}$$

This method of obtaining a limiting solution is to be preferred in the case of arbitrary boundary and variable coefficients. In fact, we proved the convergence of the half sum of fractional steps without excluding any fractional steps and without requiring the difference operators to be commutative, a property which is not satisfied in the case of variable coefficients and an arbitrary boundary. G. Birkhoff and R. S. Varga mentioned in their work that the commutative property is an important element in the proof of convergence [56]. Their argument is invalid when the above method of passage to the limit is used. It is evident that this algorithm is applicable for arbitrary operators Λ_1, Λ_2. In the case of Laplace's equation it is indicated in [57].

4.10 Reduction of schemes of incomplete approximation to schemes of complete approximation

The analysis of boundary conditions from Sec. 2.9 is fully applicable to iterative schemes. A worsening of the approximation near the boundary leads to a situation that, even if the scheme in whole steps has the property of complete consistency, the scheme in fractional steps may not have the same properties.

Consider, for example, a scheme in whole steps

$$\left.\begin{aligned} &\left(E - \frac{\tau}{2}\Lambda_1\right)\left(E - \frac{\tau}{2}\Lambda_2\right) u^{n+1} = \left(E + \frac{\tau}{2}\Lambda_1\right)\left(E + \frac{\tau}{2}\Lambda_2\right) u^n, \\ &\qquad (x_1, x_2) \in G\colon \{0 < x_i < \pi\}, \qquad i = 1, 2; \\ &u^n(x_1, x_2) = u^{n+1}(x_1, x_2) = f(x_1, x_2), \qquad (x_1, x_2) \in \gamma. \end{aligned}\right\} \tag{4.10.1}$$

The splitting scheme

$$\begin{aligned} \frac{u^{n+1/2} - u^n}{\tau} &= \Lambda_1 \frac{u^n + u^{n+1/2}}{2}; \\ \frac{u^{n+1} - u^{n+1/2}}{\tau} &= \Lambda_2 \frac{u^{n+1/2} + u^{n+1}}{2}, \end{aligned} \tag{4.10.2}$$

for which the following boundary conditions apply

$$u^n(x_1, x_2) = u^{n+1/2}(x_1, x_2) = u^{n+1}(x_1, x_2) = f(x_1, x_2), \qquad (x_1, x_2) \in \gamma, \tag{4.10.3}$$

does not have the property of complete consistency because the scheme equivalent to it, considering boundary conditions (4.10.3), is

$$\left(E - \frac{\tau}{2}\Lambda_1\right)\left(E - \frac{\tau}{2}\Lambda_2\right) u^{n+1} = \left(E + \frac{\tau}{2}\Lambda_1\right)\left(E + \frac{\tau}{2}\Lambda_2\right) u^n + R, \tag{4.10.4}$$

where $R \neq 0$ on ω (see Sec. 2.9).

Using the method of undetermined functions and assuming that

$$\left.\begin{aligned}\frac{u^{n+1/2}-u^n}{\tau} &= \Lambda_1 \frac{u^n+u^{n+1/2}}{2}+q_1;\\ \frac{u^{n+1}-u^{n+1/2}}{\tau} &= \Lambda_2 \frac{u^{n+1/2}+u^{n+1}}{2}+q_2,\end{aligned}\right. \tag{4.10.5}$$

it is possible to select q_1, q_2 in such a way as to make $R = 0$ and to satisfy boundary conditions (4.10.3). After this the scheme (4.10.5) becomes the scheme of complete approximation. An analogous approach is used in the case when the scheme does not possess complete consistency because of the structure of difference operator and not merely because of the boundary conditions. Let us demonstrate this in the splitting scheme. It is known that the splitting scheme in the three-dimensional case does not possess the property of complete consistency. Consider a splitting scheme

$$\left.\begin{aligned}\frac{u^{n+1/3}-u^n}{\tau} &= \Lambda_1(\alpha\, u^{n+1/3}+\beta\, u^n)+q_1;\\ \frac{u^{n+2/3}-u^{n+1/3}}{\tau} &= \Lambda_2(\alpha\, u^{n+2/3}+\beta\, u^{n+1/3})+q_2;\\ \frac{u^{n+1}-u^{n+2/3}}{\tau} &= \Lambda_3(\alpha\, u^{n+1}+\beta\, u^{n+2/3})+q_3\end{aligned}\right\} \tag{4.10.6}$$

with undetermined right-hand sides q_1, q_2 and q_3. A corresponding scheme in whole steps is

$$\begin{aligned}&(E-\alpha\tau\Lambda_1)(E-\alpha\tau\Lambda_2)(E-\alpha\tau\Lambda_3)\,u^{n+1}\\ &\qquad = (E+\beta\tau\Lambda_1)(E+\beta\tau\Lambda_2)(E+\beta\tau\Lambda_3)\,u^n+\tau Q,\end{aligned} \tag{4.10.7}$$

where

$$\begin{aligned}Q = {}&(E+\beta\tau\Lambda_2)(E+\beta\tau\Lambda_3)\,q_1+(E-\alpha\tau\Lambda_1)(E+\beta\tau\Lambda_3)\,q_2+\\ &+(E-\alpha\tau\Lambda_1)(E-\alpha\tau\Lambda_2)\,q_3.\end{aligned}$$

Scheme (4.10.7) can be transformed into

$$(E-\alpha\tau\Lambda_1)(E-\alpha\tau\Lambda_2)(E-\alpha\tau\Lambda_3)\frac{u^{n+1}-u^n}{\tau} = \Lambda\, u^n+R, \tag{4.10.8}$$

where

$$\begin{aligned}R = {}&[\tau(\beta^2-\alpha^2)(\Lambda_1\Lambda_2+\Lambda_1\Lambda_3+\Lambda_2\Lambda_3)+\\ &+\tau^2(\beta^3+\alpha^3)\Lambda_1\Lambda_2\Lambda_3]\,u^n+Q = \Phi\, u^n+Q.\end{aligned} \tag{4.10.9}$$

We can require now that scheme (4.10.6) be the scheme of complete consistency. For this to be true it is necessary and sufficient that $R = 0$. If we assume that $q_1 = q_2 = 0$ inside region G, then for q_3 we have the following equation

$$(E-\alpha\tau\Lambda_1)(E-\alpha\tau\Lambda_2)\,q_3 = -\Phi\, u^n. \tag{4.10.10}$$

By assigning q_1, q_2 and q_3 on the boundary to correspond with the boundary conditions, q_3 is determined from Eq. (4.10.10). When q_1, q_2 and q_3 are selected as above the scheme (4.10.6) becomes a scheme of complete consistency.

Chapter 5

Boundary Value Problems in the Theory of Elasticity

5.1 The equation of elastic equilibrium and elastic vibrations

The deformation of a plane elastic flat body is characterized by a tensor (deformation tensor):

$$\varepsilon_{ij} = \frac{1}{2}\left(\frac{\partial u_i}{\partial x_j} + \frac{\partial u_j}{\partial x_i}\right), \qquad i, j = 1, 2. \tag{5.1.1}$$

Stresses in a body caused by the deformation are characterized by the tensor σ_{ij} (stress tensor). According to Hooke's law, the tensors σ_{ij} and ε_{ij} are related by the linear relation

$$\sigma_{ij} = \lambda\, \delta_{ij}\, \varepsilon + 2\mu\, \varepsilon_{ij}, \qquad \varepsilon = \varepsilon_{11} + \varepsilon_{22} = \operatorname{div} \boldsymbol{u}, \tag{5.1.2}$$

where $\delta_{ij} = \begin{cases} 0, & i \neq j, \\ 1, & i = j, \end{cases}$ λ, μ are Lamé coefficients.

The conditions of elastic equilibrium are

$$\begin{aligned} \frac{\partial \sigma_{11}}{\partial x_1} + \frac{\partial \sigma_{12}}{\partial x_2} + \varrho X_1 &= 0; \\ \frac{\partial \sigma_{21}}{\partial x_1} + \frac{\partial \sigma_{22}}{\partial x_2} + \varrho X_2 &= 0, \end{aligned} \tag{5.1.3}$$

where X_1 and X_2 are the components of body forces.

According to the D'Alembert principle the equations of elastic vibrations are

$$\begin{aligned} -\varrho \frac{\partial^2 u_1}{\partial t^2} + \frac{\partial \sigma_{11}}{\partial x_1} + \frac{\partial \sigma_{12}}{\partial x_2} + \varrho X_1 &= 0; \\ -\varrho \frac{\partial^2 u_2}{\partial t^2} + \frac{\partial \sigma_{21}}{\partial x_1} + \frac{\partial \sigma_{22}}{\partial x_2} + \varrho X_2 &= 0, \end{aligned} \tag{5.1.4}$$

where ϱ is the density of the material.

Using Eqs. (5.1.1) and (5.1.2), Eq. (5.1.4) can be represented as follows

$$-\varrho \frac{\partial^2 \boldsymbol{u}}{\partial t^2} + (\lambda + \mu) \operatorname{grad} \operatorname{div} \boldsymbol{u} + \mu \Delta \boldsymbol{u} + \varrho \boldsymbol{X} = 0. \tag{5.1.5}$$

Eqs. (5.1.3) can be written in a corresponding form.

Eq. (5.1.5) can be changed somewhat

$$-\varrho \frac{\partial^2 u_1}{\partial t^2} + (\lambda + 2\mu) \frac{\partial^2 u_1}{\partial x_1^2} + \mu \frac{\partial^2 u_1}{\partial x_2^2} + (\lambda + \mu) \frac{\partial^2 u_2}{\partial x_1 \partial x_2} + \varrho X_1 = 0;$$
$$-\varrho \frac{\partial^2 u_2}{\partial t^2} + (\lambda + 2\mu) \frac{\partial^2 u_2}{\partial x_2^2} + \mu \frac{\partial^2 u_2}{\partial x_1^2} + (\lambda + \mu) \frac{\partial^2 u_1}{\partial x_1 \partial x_2} + \varrho X_2 = 0. \qquad (5.1.6)$$

Eqs. (5.1.3) can be changed analogously.

In the absence of body forces ($X_1 = X_2 = 0$), assuming that

$$\sigma_{11} = \frac{\partial^2 \psi}{\partial x_2^2}, \quad \sigma_{22} = \frac{\partial^2 \psi}{\partial x_1^2}; \quad \sigma_{12} = -\frac{\partial^2 \psi}{\partial x_1 \partial x_2}, \qquad (5.1.7)$$

expression (5.1.3) becomes an identity. The function ψ is not arbitrary. Eq. (5.1.7) represent an overdetermined system of equations for u_i.

The compatibility condition of this system has the form

$$\Delta \Delta \psi = 0. \qquad (5.1.8)$$

In fact, if we introduce the quantity

$$\omega = \frac{1}{2} \left(\frac{\partial u_2}{\partial x_1} - \frac{\partial u_1}{\partial x_2} \right), \qquad (5.1.9)$$

we have

$$\frac{\partial u_1}{\partial x_1} = \varepsilon_{11}; \qquad \frac{\partial u_1}{\partial x_2} = \varepsilon_{12} - \omega; \qquad (5.1.10)$$

$$\frac{\partial u_2}{\partial x_1} = \varepsilon_{12} + \omega; \qquad \frac{\partial u_2}{\partial x_2} = \varepsilon_{22}. \qquad (5.1.11)$$

The compatibility conditions (5.1.10), (5.1.11) give

$$\frac{\partial \omega}{\partial x_1} = \frac{\partial \varepsilon_{12}}{\partial x_1} - \frac{\partial \varepsilon_{11}}{\partial x_2} \qquad (5.1.12)$$

and

$$\frac{\partial \omega}{\partial x_2} = \frac{\partial \varepsilon_{22}}{\partial x_1} - \frac{\partial \varepsilon_{12}}{\partial x_2}, \qquad (5.1.13)$$

respectively. Finally, the compatibility conditions (5.1.12) and (5.1.13) give

$$\frac{\partial}{\partial x_2} \left(\frac{\partial \varepsilon_{12}}{\partial x_1} - \frac{\partial \varepsilon_{11}}{\partial x_2} \right) - \frac{\partial}{\partial x_1} \left(\frac{\partial \varepsilon_{22}}{\partial x_1} - \frac{\partial \varepsilon_{12}}{\partial x_2} \right)$$
$$= - \left[\frac{\partial^2 \varepsilon_{11}}{\partial x_2^2} + \frac{\partial^2 \varepsilon_{22}}{\partial x_1^2} - 2 \frac{\partial^2 \varepsilon_{12}}{\partial x_1 \partial x_2} \right] = 0. \qquad (5.1.14)$$

Using Eqs. (5.1.2) and (5.1.7), condition (5.1.14) leads to Eq. (5.1.8).

Harmonic analysis of the stability of Eq. (5.1.5) shows that elastic vibrations do not decay. Consequently, the solution of Eq. (5.1.3) cannot be obtained from the solution of Eq. (5.1.4) by transition to the steady regime. The decay of elastic vibrations is described by the following equation

$$\alpha \frac{\partial \boldsymbol{u}}{\partial t} + \beta \varrho \frac{\partial^2 \boldsymbol{u}}{\partial t^2} = (\lambda + \mu) \operatorname{grad} \operatorname{div} \boldsymbol{u} + \mu \Delta \boldsymbol{u} + \varrho \boldsymbol{X}, \qquad (5.1.15)$$
$$\alpha > 0, \quad \beta > 0.$$

eqn's of elastic vibration with ("viscous") decay

Eq. (5.1.15) enables us to obtain iterative schemes for the solution of the steady equation. In order to obtain the simplest iterative schemes the equation of a purely parabolic type can be used:

$$\alpha \frac{\partial \boldsymbol{u}}{\partial t} = (\lambda + \mu) \operatorname{grad} \operatorname{div} \boldsymbol{u} + \mu \Delta \boldsymbol{u} + \varrho \boldsymbol{X}. \tag{5.1.16}$$

An analogous approach can be used for Eq. (5.1.8). We use the corresponding equation

$$\alpha \frac{\partial \psi}{\partial t} + \beta \frac{\partial^2 \psi}{\partial t^2} + \Delta \Delta \psi = 0, \quad \alpha > 0, \quad \beta > 0, \tag{5.1.17}$$

or simply the equation

$$\alpha \frac{\partial \psi}{\partial t} + \Delta \Delta \psi = 0. \tag{5.1.18}$$

5.2 Boundary value problems in the theory of elasticity

The following problems may be posed for Eq. (5.1.5).

(i) First boundary value problem. The displacements u_1, u_2 are given as functions of s (the arc parameter along γ) and of t on the boundary γ of a plane region G:

$$\boldsymbol{u}(s) = \boldsymbol{f}(s, t). \tag{5.2.1}$$

(ii) Second boundary value problem. Normal and tangential stresses are given on the boundary γ:

$$\sigma_n(s) = f_1(s, t); \quad \sigma_\tau(s) = f_2(s, t). \tag{5.2.2}$$

In the case of the steady problem the functions $f_1(s)$ and $f_2(s)$ must satisfy an additional condition, namely, that of equilibrium of the body regarded as a solid.

(iii) Boundary value problem of mixed type. Displacements are given on one section of the boundary and stresses on the other.

(iv) In the case of the biharmonic equation (5.1.8) or the corresponding inhomogeneous equation

$$\Delta \Delta \psi + q = 0 \tag{5.2.3}$$

the boundary value problem of the following type is considered:

$$\psi = 0, \quad \Delta \psi = f(s). \tag{5.2.4}$$

We arrive at Eqs. (5.2.3) and (5.2.4) in the case, for example, of a freely supported plate. In this case ψ is considered as the displacement of the median plane of the plate; q is the load; $f(s) = 0$ (see [58]).

5.3 The integration scheme for the unsteady equations of elasticity

The explicit scheme of second order accuracy

$$\left.\begin{aligned} -\frac{u_i^{n+1}-2u_i^n+u_i^{n-1}}{\tau^2}+(\lambda+2\mu)\frac{\Delta_i\Delta_{-i}}{h_i^2}u_i^n+\mu\frac{\Delta_{3-i}\Delta_{-3+i}}{h_{3-i}^2}u_i^n+ \\ +(\lambda+\mu)\frac{(\Delta_1+\Delta_{-1})(\Delta_2+\Delta_{-2})}{4h_1h_2}u_{3-i}^n+X_i^n=0, \\ i=1,2 \end{aligned}\right\} \quad (5.3.1)$$

is used for Eqs. (5.1.5) and (5.1.6) with $\varrho = 1$. A. N. Konovalov [59] suggested an implicit scheme of second order accuracy for Eqs. (5.1.5) and (5.1.6), which is based on an approximate factorization of an operator. Consider the scheme

$$\left.\begin{aligned} &-\frac{u_i^{n+1}-2u_i^n+u_i^{n-1}}{\tau^2}+(\lambda+2\mu)\frac{\Delta_i\Delta_{-i}}{h_i^2}\frac{u_i^{n+1}+u_i^{n-1}}{2}+ \\ &+\mu\frac{\Delta_{3-i}\Delta_{-3+i}}{h_{3-i}^2}\frac{u_i^{n+1}+u_i^{n-1}}{2}+ \\ &+(\lambda+\mu)\frac{(\Delta_1+\Delta_{-1})(\Delta_2+\Delta_{-2})}{4h_1h_2}u_{3-i}^n+X_i^n \\ &=-\frac{u_i^{n+1}-2u_i^n+u_i^{n-1}}{\tau^2}+\frac{\lambda+2\mu}{2}\Lambda_{ii}(u_i^{n+1}+u_i^{n-1})+ \\ &+\frac{\mu}{2}\Lambda_{3-i,3-i}(u_i^{n+1}+u_i^{n-1})+(\lambda+\mu)\Lambda_{12}u_{3-i}^n+X_i^n=0, \\ &\qquad i=1,2, \end{aligned}\right\} \quad (5.3.2)$$

where

$$\Lambda_{ii}=\frac{\Delta_i\Delta_{-i}}{h_i^2}; \qquad \Lambda_{12}=\frac{(\Delta_1+\Delta_{-1})(\Delta_2+\Delta_{-2})}{4h_1h_2}.$$

Scheme (5.3.2) is of second order accuracy, absolutely stable, but is difficult to use. Factorization of the upper operator of Eq. (5.3.2) results in

$$\begin{aligned} &\left[E-\frac{\tau^2}{2}(\lambda+2\mu)\Lambda_{ii}\right]\left[E-\frac{\tau^2}{2}\mu\Lambda_{3-i,3-i}\right](u_i^{n+1}+u_i^{n-1}) \\ &\qquad =2u_i^n+(\lambda+\mu)\tau^2\Lambda_{12}u_{3-i}^n+X_i^n\tau^2, \quad i=1,2. \end{aligned} \quad (5.3.3)$$

Scheme (5.3.3) is of second order accuracy, is absolutely stable, and is carried out with three-point sweeps.

A. N. Konovalov proved the convergence of scheme (5.3.3) for the first boundary value problem by the method of energy inequalities.

5.4 Iterative schemes of solution of boundary value problems for the biharmonic equation

As above we consider iterative schemes for the solution of the equation

$$\Delta\Delta\psi = 0 \quad (5.4.1)$$

as integration schemes of the equation

$$\frac{\partial \psi}{\partial t} + \Delta\Delta\psi = \frac{\partial \psi}{\partial t} + (L_{11} + L_{12} + L_{21} + L_{22})\,\psi = 0;$$
$$L_{ij} = \frac{\partial^4}{\partial x_i^2\,\partial x_j^2}, \qquad i, j = 1, 2. \tag{5.4.2}$$

(i) *Splitting schemes.* Let us construct this scheme like that for the equation of heat conduction of general type (see Sec. 2.4)

$$\frac{\psi^{n+1/2} - \psi^n}{\tau} + \Lambda_{11}\,\psi^{n+1/2} + \Lambda_{12}\,\psi^n = 0;$$
$$\frac{\psi^{n+1} - \psi^{n+1/2}}{\tau} + \Lambda_{21}\,\psi^{n+1/2} + \Lambda_{22}\,\psi^{n+1} = 0, \tag{5.4.3}$$

where

$$\Lambda_{11} = \left(\frac{\Delta_1\Delta_{-1}}{h_1^2}\right)^2; \qquad \Lambda_{22} = \left(\frac{\Delta_2\Delta_{-2}}{h_2^2}\right)^2; \qquad \Lambda_{12} = \Lambda_{21} = \frac{\Delta_1\Delta_{-1}\Delta_2\Delta_{-2}}{(h_1 h_2)^2}. \tag{5.4.4}$$

The corresponding scheme in whole steps is

$$(E + \tau\Lambda_{11})\,(E + \tau\Lambda_{22})\,\psi^{n+1} = (E - \tau\Lambda_{12})^2\,\psi^n. \tag{5.4.5}$$

From this it follows that the scheme is strongly stable, since

$$\varrho = \frac{(1 - a_1 a_2)^2}{(1 + a_1^2)\,(1 + a_2^2)}; \qquad a_i = -\frac{4\sqrt{\tau}\,\sin^2\dfrac{k_i h_i}{2}}{h_i^2}. \tag{5.4.6}$$

Taking into account the identity

$$\Lambda_{11}\Lambda_{22} = \Lambda_{12}^2, \tag{5.4.7}$$

we see that the splitting scheme (5.4.3) for the biharmonic equation, as opposed to the equation of heat conduction, possesses the property of complete consistency.

(ii) *The scheme of stabilizing correction.* S. D. Conte and R. J. Dames [60] generalized the scheme of stabilizing correction of Douglas and Rachford for the biharmonic equation

$$\frac{\psi^{n+1/2} - \psi^n}{\tau} + \Lambda_{11}\,\psi^{n+1/2} + 2\Lambda_{12}\,\psi^n + \Lambda_{22}\,\psi^n = 0;$$
$$\frac{\psi^{n+1} - \psi^{n+1/2}}{\tau} + \Lambda_{22}(\psi^{n+1} - \psi^n) = 0. \tag{5.4.8}$$

The corresponding scheme in whole steps is

$$\frac{\psi^{n+1} - \psi^n}{\tau} + (\Lambda_{11} + \Lambda_{22})\,\psi^{n+1} + 2\Lambda_{12}\,\psi^n + \tau\Lambda_{11}\Lambda_{22}(\psi^{n+1} - \psi^n) = 0. \tag{5.4.8'}$$

Because of Eq. (5.4.7) scheme (5.4.8) is equivalent to scheme (5.4.5).

(iii) *The scheme of approximation corrections*

$$\frac{\psi^{n+1/4}-\psi^n}{\alpha\tau}+\Lambda_{11}\psi^{n+1/4}=0;\quad \frac{\psi^{n+1/2}-\psi^{n+1/4}}{\alpha\tau}+\Lambda_{22}\psi^{n+1/2}=0;$$
$$\frac{\psi^{n+1}-\psi^n}{\tau}+\Lambda\psi^{n+1/2}=0;\quad \Lambda=\Lambda_{11}+2\Lambda_{12}+\Lambda_{22}. \qquad (5.4.9)$$

The scheme in whole steps is

$$(E+\alpha\tau\Lambda_{11})(E+\alpha\tau\Lambda_{22})\frac{\psi^{n+1}-\psi^n}{\tau}+\Lambda\psi^n=0 \qquad (5.4.10)$$

and is equivalent to scheme (5.4.5) for $\alpha = 1$.

(iv) *The scheme of stabilizing operator*

$$A\frac{\psi^{n+1}-\psi^n}{\tau}+\Lambda\psi^n=0;\quad \Lambda=\Lambda_{11}+\Lambda_{22}+2\Lambda_{12};$$
$$A=(E+\alpha\tau\Lambda_{11})(E+\alpha\tau\Lambda_{22}) \qquad (5.4.11)$$

is also equivalent to scheme (5.4.5) for $\alpha = 1$.

Thus, schemes of splitting, corrections and stabilizing operator for $\alpha = 1$ represent different realizations of the same uniform scheme. In addition, they differ in the satisfying of boundary conditions. Note that schemes of splitting and stabilizing operator are two-layered, and schemes of corrections are three-layered.

(v) *The splitting scheme for the system of harmonic equations.* For some boundary value problems of elasticity (for example, for the transverse bending of a freely loaded supported plate) it is more convenient to use a system of two harmonic equations instead of the biharmonic equation.

For the boundary value problem

$$\Delta\psi=\varphi;\quad \Delta\varphi+q=0; \qquad (5.4.12)$$
$$\varphi(s)=f(s);\quad \psi(s)=g(s) \qquad (5.4.13)$$

consider the splitting scheme

$$\left.\begin{aligned}
\frac{\varphi^{n+1/2}-\varphi^n}{\tau}&=\Lambda_1(\alpha\varphi^{n+1/2}+\beta\varphi^n);\\
\frac{\varphi^{n+1}-\varphi^{n+1/2}}{\tau}&=\Lambda_2(\alpha\varphi^{n+1}+\beta\varphi^{n+1/2})+q;\\
\frac{\psi^{n+1/2}-\psi^n}{\tau}&=\Lambda_1(\alpha\psi^{n+1/2}+\beta\psi^n);\\
\frac{\psi^{n+1}-\psi^{n+1/2}}{\tau}&=\Lambda_2(\alpha\psi^{n+1}+\beta\psi^{n+1/2})+\varphi^{n+1}.
\end{aligned}\right\} \qquad (5.4.14)$$

Scheme (5.4.14) is strongly stable and is completely consistent for $\alpha = 1/2$.

For passage to the limit the half-sum of the values of the function at whole and fractional steps should be taken (see Sec. 4.9).

5.5 Iterative schemes for the system of equations of elastic displacements

For the equations of elastic equilibrium in displacements

$$(\lambda + 2\mu)\frac{\partial^2 u_1}{\partial x_1^2} + \mu\frac{\partial^2 u_1}{\partial x_2^2} + (\lambda + \mu)\frac{\partial^2 u_2}{\partial x_1\,\partial x_2} = 0;$$
$$(\lambda + \mu)\frac{\partial^2 u_1}{\partial x_1\,\partial x_2} + \mu\frac{\partial^2 u_2}{\partial x_1^2} + (\lambda + 2\mu)\frac{\partial^2 u_2}{\partial x_2^2} = 0. \qquad (5.5.1)$$

A. N. Konovalov [61] suggested the scheme of stabilizing correction

$$\left.\begin{aligned}
\frac{u_1^{n+1/2} - u_1^n}{\tau} &= \Lambda_{11}\, u_1^{n+1/2} + \Lambda_{12}\, u_1^n + \Omega\, u_2^n,\\
\frac{u_2^{n+1/2} - u_2^n}{\tau} &= \Omega\, u_1^n + \Lambda_{21}\, u_2^{n+1/2} + \Lambda_{22}\, u_2^n,\\
\frac{u_1^{n+1} - u_1^{n+1/2}}{\tau} &= \Lambda_{12}(u_1^{n+1} - u_1^n),\\
\frac{u_2^{n+1} - u_2^{n+1/2}}{\tau} &= \Lambda_{22}(u_2^{n+1} - u_2^n),
\end{aligned}\right\} \qquad (5.5.2)$$

where

$$\Lambda_{11} = (\lambda + 2\mu)\frac{\Delta_1\Delta_{-1}}{h_1^2}; \qquad \Lambda_{12} = \mu\frac{\Delta_2\Delta_{-2}}{h_2^2}; \qquad \Lambda_{21} = \mu\frac{\Delta_1\Delta_{-1}}{h_1^2};$$
$$\Lambda_{22} = (\lambda + 2\mu)\frac{\Delta_2\Delta_{-2}}{h_2^2}; \qquad \Omega = (\lambda + \mu)\frac{\Delta_1 + \Delta_{-1}}{2h_1}\,\frac{\Delta_2 + \Delta_{-2}}{2h_2}.$$

A. N. Konovalov proved the convergence of this scheme for the first boundary value problem in a rectangle. A. A. Samarskii also suggested several schemes with fractional steps for the equations of elasticity [96].

5.6 Boundary conditions in problems of elasticity

The solution of boundary value problems of the second type requires construction of recurrent schemes to take account of boundary conditions. The boundary value problem presented below demonstrates this (Fig. 5).

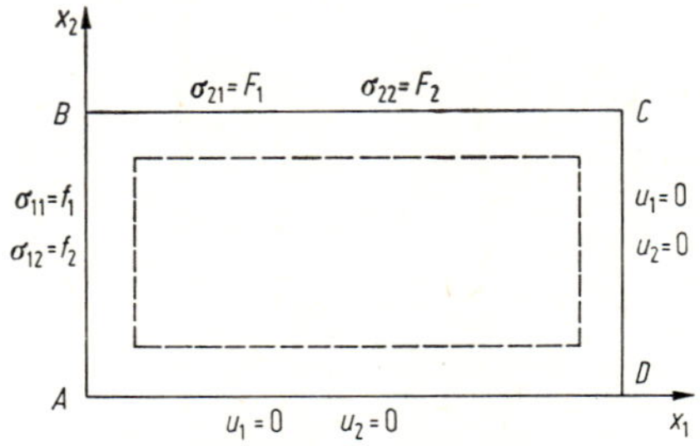

Fig. 5. Solution of a mixed boundary value problem for the equation of elasticity in a rectangle

The boundary conditions at the left vertical boundary are

$$\mu \frac{\partial u_1}{\partial x_2} + \mu \frac{\partial u_2}{\partial x_1} = f_2; \qquad \lambda \frac{\partial u_2}{\partial x_2} + (\lambda + 2\mu) \frac{\partial u_1}{\partial x_1} = f_1. \tag{5.6.1}$$

System (5.6.1) is of hyperbolic type in which, in the case of vertical boundaries, x_2 is the time variable, and x_1 is the space variable. Quantities u_1 and u_2 at the boundary are denoted by u_1^n and u_2^n; U_1^n and U_2^n are values at the internal network points close to the boundary. The simplest explicit approximation of Eq. (5.6.1) results in

$$\mu \frac{u_1^{n+1} - u_1^n}{h_2} - \mu \frac{u_2^n - U_2^n}{h_1} = f_2^n; \qquad \lambda \frac{u_2^{n+1} - u_2^n}{h_2} - (\lambda + 2\mu) \frac{u_1^n - U_1^n}{h_1} = f_1^n. \tag{5.6.2}$$

To test the stability of Eq. (5.6.2) for fixed U_1^n, U_2^n, f_1^n and f_2^n an equation is derived in variations of the unknowns (the variation sign is omitted here)

$$\frac{u_1^{n+1} - u_1^n}{h_2} - \frac{u_2^n}{h_1} = 0, \qquad \lambda \frac{u_2^{n+1} - u_2^n}{h_2} - (\lambda + 2\mu) \frac{u_1^n}{h_1} = 0. \tag{5.6.3}$$

It follows from this that

$$\boldsymbol{u}^{n+1} = C\, \boldsymbol{u}^n; \qquad \boldsymbol{u} = \{u_1, u_2\}; \qquad C = \left\| \begin{matrix} 1 & \dfrac{h_2}{h_1} \\ \dfrac{\lambda + 2\mu}{\lambda} \cdot \dfrac{h_2}{h_1} & 1 \end{matrix} \right\|. \tag{5.6.3'}$$

We now evaluate the norm of the transition matrix C from Eq. (5.6.3′). For its characteristic roots we have

$$\varrho_{1,2} = 1 \pm \sqrt{\frac{\lambda + 2\mu}{\lambda} \cdot \frac{h_2}{h_1}}. \tag{5.6.4}$$

It is evident that the spectral radius and the norm of the matrix C are larger than 1 and the recurrent calculation (5.6.2) is unstable.

Let us apply the implicit approximation

$$u \frac{u_1^{n+1} - u_1^n}{h_2} - \mu \frac{u_2^{n+1} - U_2^{n+1}}{h_1} = f_2^n; \qquad \lambda \frac{u_2^{n+1} - u_2^n}{h_2} - (\lambda + 2\mu) \frac{u_1^{n+1} - U_1^{n+1}}{h_1} = f_1^n. \tag{5.6.5}$$

The equations in variations are

$$\frac{u_1^{n+1} - u_1^n}{h_2} - \frac{u_2^{n+1}}{h_1} = 0; \qquad \lambda \frac{u_2^{n+1} - u_2^n}{h_2} - (\lambda + 2\mu) \frac{u_1^{n+1}}{h_1} = 0. \tag{5.6.6}$$

From these we have

$$u_1^n = u_1^{n+1} - \frac{h_2}{h_1} u_2^{n+1}; \qquad u_2^n = -\frac{\lambda + 2\mu}{\lambda} \cdot \frac{h_2}{h_1} u_1^{n+1} + u_2^{n+1}. \tag{5.6.7}$$

For the calculation to be stable it is necessary that the matrix (inverse of the transformation matrix)

$$\begin{Vmatrix} 1 & -\frac{h_2}{h_1} \\ -\frac{\lambda + 2\mu}{\lambda} \cdot \frac{h_2}{h_1} & 1 \end{Vmatrix} \tag{5.6.8}$$

has characteristic roots ≥ 1. For the roots ϱ of the matrix we have

$$\varrho_{1,2} = 1 \pm \sqrt{\frac{\lambda + 2\mu}{\lambda}} \cdot \frac{h_2}{h_1}. \tag{5.6.9}$$

The stability condition is therefore

$$\sqrt{\frac{\lambda + 2\mu}{\lambda}} \cdot \frac{h_2}{h_1} \geq 2. \tag{5.6.10}$$

If an analogous scheme is written for the upper horizontal edge, h_1 and h_2 change their roles and we have

$$\sqrt{\frac{\lambda + 2\mu}{\lambda}} \cdot \frac{h_1}{h_2} \geq 2. \tag{5.6.11}$$

As a result the calculation will be unstable on one of the sides. In order to achieve a stable recurrent calculation on the horizontal side, when the stability conditions are satisfied on the vertical side, it is necessary either to enlarge the network along x_1, or to make it smaller along x_2.

Still another calculation scheme of the boundary can be indicated, which is based on running calculation.

Boundary conditions of the second type are presented in this case at a local reference point, where one vector is directed along tangent to the boundary clockwise, and the other along the external normal (Fig. 6).

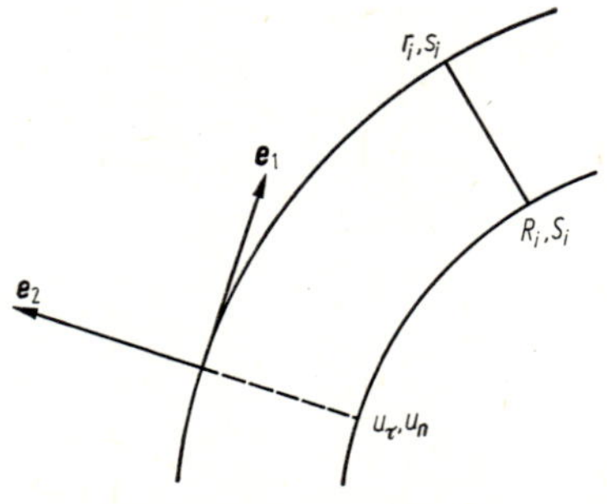

Fig. 6. Boundary conditions of the second type at the local reference point

Then the displacement vector $\boldsymbol{u}$ is

$$\boldsymbol{u} = u_\tau \mathbf{e}_1 + u_n \mathbf{e}_2. \tag{5.6.12}$$

Contrary to the case of a rectangle, the direction $\mathbf{e}_2$ is considered to be in time; $\mathbf{e}_1$ is the space direction; local Cartesian coordinates are represented by t and x, respectively.

Then the boundary conditions are

$$\begin{aligned} (\lambda + 2\mu)\frac{\partial u_n}{\partial t} + \lambda\frac{\partial u_\tau}{\partial x} &= f_1; \\ \mu\left(\frac{\partial u_\tau}{\partial t} + \frac{\partial u_n}{\partial x}\right) &= f_2. \end{aligned} \tag{5.6.13}$$

Let us write Eq. (5.6.13) in invariant form

$$\frac{\partial r}{\partial t} + c\frac{\partial r}{\partial x} = g_1; \qquad \frac{\partial s}{\partial t} - c\frac{\partial s}{\partial x} = g_2, \tag{5.6.14}$$

where

$$\begin{aligned} r = u_n + c\,u_\tau; \quad s = u_n - c\,u_\tau, \quad c = \sqrt{\frac{\lambda}{\lambda + 2\mu}}; \\ g_1 = \frac{f_1}{\lambda + 2\mu} + c\frac{f_2}{\mu}; \quad g_2 = \frac{f_1}{\lambda + 2\mu} - c\frac{f_2}{\mu}. \end{aligned} \tag{5.6.15}$$

Eqs. (5.6.14) are then solved with the aid of the implicit majorant scheme

$$\frac{r_i - R_i}{\tau} + c\frac{r_i - r_{i-1}}{h} = g_{1\,i}; \qquad \frac{s_i - S_i}{\tau} - c\frac{s_{i+1} - s_i}{h} = g_{2\,i}, \tag{5.6.16}$$

which is stable for any τ and h.

In the case of a rectangular region (Fig. 5) on the side AB we have

$$\begin{aligned} u_1 = -u_n; \quad u_2 = u_\tau; \quad t = -x_1; \quad x = x_2; \quad \tau = h_1; \quad h = h_2; \\ r = -u_1 + c u_2; \quad s = -u_1 - c u_2, \end{aligned} \tag{5.6.17}$$

and on the side BC

$$\begin{aligned} u_1 = u_\tau; \quad u_2 = u_n; \quad t = x_2; \quad x = x_1; \quad \tau = h_2; \quad h = h_1; \\ r = u_2 + c u_1; \quad s = u_2 - c u_1. \end{aligned} \tag{5.6.18}$$

The transition from the invariants (5.6.17) to those of Eq. (5.6.18) is accomplished along some arc which includes the angle ABC.

In conclusion it should be noted that only a few problems and methods have been studied in the area of boundary value problems of elasticity.

Chapter 6

Schemes of Higher Accuracy

Up to this point we have considered difference schemes of integration with accuracy of $O(\tau^\alpha + h^\beta)$, $\alpha, \beta \leq 2$. The corresponding iterative schemes produce accuracy of $O(h^\beta)$. The method of fractional steps can be applied to obtain simple schemes of higher accuracy (s.h.a.), $\beta > 2$.

6.1 Uniform schemes of higher accuracy

Following the works of J. Douglas and J. E. Gunn [62], and of A. A. Samarskii [63] we consider now several uniform s.h.a. for the equation of heat conduction

$$\frac{\partial u}{\partial t} = \sum_{i=1}^{m} \frac{\partial^2 u}{\partial x_i^2}, \tag{6.1.1}$$

$$\frac{\partial u}{\partial t} = \sum_{i,j=1}^{m} a_{ij} \frac{\partial^2 u}{\partial x_i \partial x_j}, \qquad a_{ij} = \text{const.} \tag{6.1.2}$$

For an arbitrary, sufficiently smooth function $u(x_1, \ldots, x_m, t)$ the following approximations are valid (it is assumed that $h_1 = h_2 = \cdots = h_m = h$):

$$\begin{aligned} \frac{\partial^2 u^n}{\partial x_i^2} &= \Lambda_i u^n - \frac{h^2}{12} \frac{\partial^4 u^n}{\partial x_i^4} + O(h^4), \\ \Delta u^n &= \Lambda u^n - \frac{h^2}{12} \sum_{i=1}^{m} \frac{\partial^4 u^n}{\partial x_i^4} + O(h^4); \end{aligned} \tag{6.1.3}$$

$$\begin{aligned} \frac{\partial^2 u^n}{\partial x_i^2} &= \frac{1}{3} \Lambda_i (u^{n-1} + u^n + u^{n+1}) - \frac{h^2}{12} \frac{\partial^4 u^n}{\partial x_i^4} + O(\tau^2 + h^4), \\ \Delta u &= \frac{1}{3} \Lambda (u^{n-1} + u^n + u^{n+1}) - \frac{h^2}{12} \sum_{i=1}^{m} \frac{\partial^4 u^n}{\partial x_i^4} + O(\tau^2 + h^4), \end{aligned} \tag{6.1.4}$$

where

$$\Delta = \sum_{i=1}^{m} \frac{\partial^2}{\partial x_i^2}; \qquad \Lambda = \sum_{i=1}^{m} \Lambda_i; \qquad u^n(x_1, \ldots, x_m) = u(x_1, \ldots, x_m, n\tau).$$

From Eq. (6.1.4) follows for the solution of Eq. (6.1.1)

$$\frac{u^{n+1}-u^{n-1}}{2\tau} = \frac{1}{3}\Lambda(u^{n-1}+u^n+u^{n+1}) - \frac{h^2}{12}\sum_{i=1}^{m}\frac{\partial^4 u^n}{\partial x_i^4} + O(\tau^2+h^4). \tag{6.1.5}$$

Since in Eq. (6.1.1)

$$\Delta^2 u = \sum_{i=1}^{m}\frac{\partial^4 u}{\partial x_i^4} + 2\sum_{i<j}\frac{\partial^4 u}{\partial x_i^2\,\partial x_j^2} = \Delta u_t = u_{tt}, \qquad i,j=1,2,\ldots,m, \tag{6.1.6}$$

then we can assume that

$$\left.\begin{aligned}\sum_{i=1}^{m}\frac{\partial^4 u^n}{\partial x_i^4} &= \frac{u^{n+1}-2u^n+u^{n-1}}{\tau^2} - 2\sum_{i<j}\frac{\partial^4 u^n}{\partial x_i^2\,\partial x_j^2} + O(\tau^2)\\ &= \frac{u^{n+1}-2u^n+u^{n-1}}{\tau^2} - 2\sum_{i<j}\Lambda_i\Lambda_j u^n + O(\tau^2+h^2),\\ &\qquad i,j=1,2,\ldots,m,\end{aligned}\right\} \tag{6.1.7}$$

or

$$\sum_{i=1}^{m}\frac{\partial^4 u^n}{\partial x_i^4} = \Lambda\frac{u^{n+1}-u^{n-1}}{2\tau} - 2\sum_{i<j}\Lambda_i\Lambda_j u^n + O(\tau^2+h^2). \tag{6.1.8}$$

Comparing Eqs. (6.1.5), (6.1.7) and (6.1.8), we obtain the following s.h.a.

$$\begin{aligned}\frac{u^{n+1}-u^{n-1}}{2\tau} &= \frac{1}{3}\Lambda(u^{n-1}+u^n+u^{n+1}) -\\ &\quad - \left(\frac{h^2}{12}\frac{u^{n+1}-2u^n+u^{n-1}}{\tau^2} - 2\sum_{i<j}\Lambda_i\Lambda_j u^n\right)\end{aligned} \tag{6.1.9}$$

or correspondingly

$$\begin{aligned}\frac{u^{n+1}-u^{n-1}}{2\tau} &= \frac{1}{3}\Lambda\left[\left(1-\frac{1}{8r}\right)u^{n+1} + u^n + \right.\\ &\quad \left.+\left(1+\frac{1}{8r}\right)u^{n-1}\right] + \frac{h^2}{6}\sum_{i<j}\Lambda_i\Lambda_j u^n,\end{aligned} \tag{6.1.10}$$

with

$$r = \tau/h^2.$$

A two-layered s.h.a. based on Eq. (6.1.3) and Eq. (6.1.6) is obtained in an analogous way:

$$\frac{u^{n+1}-u^n}{\tau} = \Lambda\frac{u^n+u^{n+1}}{2} - \frac{h^2}{12}\Lambda\frac{u^{n+1}-u^n}{\tau} + \frac{h^2}{6}\sum_{i<j}\Lambda_i\Lambda_j u^n. \tag{6.1.11}$$

Scheme (6.1.11) can also be written in the form

$$\begin{aligned}\frac{u^{n+1}-u^n}{\tau} &= \alpha\Lambda u^{n+1} + (1-\alpha)\Lambda u^n + \frac{h^2}{6}\sum_{i<j}\Lambda_i\Lambda_j u^n,\\ \alpha &= \frac{1}{2}\left(1-\frac{h^2}{6\tau}\right) = \frac{1}{2}\left(1-\frac{1}{6r}\right).\end{aligned} \tag{6.1.12}$$

Schemes (6.1.9), (6.1.10), (6.1.11) and (6.1.12) are absolutely stable and have an accuracy of order $O(\tau^2) + O(h^4)$.

6.2 Factorized schemes of higher accuracy for the equation of heat conduction

Schemes of higher accuracy have been considered in the works of J. Douglas and J. E. Gunn [62], A. A. Samarskii [63], A. A. Samarskii and V. B. Andreev [64] for parabolic equations of type (6.1.1) and (6.1.2).

These authors constructed easily realizable s.h.a. with a factorized ("split") upper operator. These schemes are constructed on the basis of m-layer uniform s.h.a. ($m \geq 2$):

$$\frac{u^{n+1} - u^n}{\tau} = A\, u^{n+1} + f^n, \tag{6.2.1}$$

where f^n is the result of application of difference space operators functions $u^n, u^{n-1}, \ldots$.

The method of construction of the factorized scheme, which was suggested by J. Douglas and J. E. Gunn [62], is as follows. Let

$$A = A_1 + \cdots + A_m \tag{6.2.2}$$

be the expansion of A into a sum of operators A_i, $i = 1, 2, \ldots, m$. Then a scheme of stabilizing correction is constructed

$$\left.\begin{aligned} &\frac{u^{n+1/m} - u^n}{\tau} = A_1(u^{n+1/m} - u^n) + A\, u^n + f^n; \\ &\frac{u^{n+2/m} - u^{n+1/m}}{\tau} = A_2\, (u^{n+2/m} - u^n); \\ &\ldots\ldots\ldots\ldots\ldots\ldots\ldots\ldots\ldots\ldots \\ &\frac{u^{n+1} - u^{n+(m-1)/m}}{\tau} = A_m(u^{n+1} - u^n). \end{aligned}\right\} \tag{6.2.3}$$

The scheme in whole steps is

$$\frac{u^{n+1} - u^n}{\tau} = A\, u^{n+1} + f^n + \Phi\left(\frac{u^{n+1} - u^n}{\tau}\right), \tag{6.2.4}$$

$$\Phi = -\tau^2 \sum_{i<j} A_i\, A_j + \tau^3 \sum_{i<j<k} A_i\, A_j\, A_k + \cdots + (-1)^{m-1}\, \tau^m A_1 \ldots A_m. \tag{6.2.5}$$

Let us consider factorized schemes which correspond to Eq. (6.1.1) ($m = 2$) and to schemes (6.1.9) and (6.1.10).

In the case of Eq. (6.1.9) we assume

$$\left.\begin{aligned} A &= \frac{2}{3}\Lambda - \frac{1}{6}\,\frac{h^2}{\tau^2}\,E; \\ A_1 &= \frac{2}{3}\Lambda_1 - \frac{1}{6}\,\frac{h^2}{\tau^2}\,E; \\ A_2 &= \frac{2}{3}\Lambda_2. \end{aligned}\right\} \tag{6.2.6}$$

Schemes (6.2.3) and (6.2.6) have order $O(\tau^2 + h^4)$ for $\tau/h = \text{const}$ and are absolutely stable.

In the case of Eq. (6.1.10) we have

$$A = \frac{2}{3}\left(1 - \frac{1}{8r}\right)\Lambda, \quad A_i = \frac{2}{3}\left(1 - \frac{1}{8r}\right)\Lambda_i, \quad i = 1, 2. \qquad (6.2.7)$$

For $\tau/h^2 =$ const schemes (6.2.3) and (6.2.7) have accuracy of $O(\tau^2 + h^4)$ and are absolutely stable.

A. A. Samarskii [63] suggested another method for the construction of a factorized operator. Scheme (6.2.1) is replaced in this case by

$$\frac{u^{n+1} - u^n}{\tau} = A\, u^{n+1} + f^n + \Phi\left(\frac{u^{n+1} - u^n}{\tau}\right)$$
$$= A\, u^{n+1} + f^n + \frac{1}{\tau}\Phi(u^{n+1} - u^n), \qquad (6.2.8)$$

where the undetermined operator Φ is selected in such a way that the upper operator in Eq. (6.2.8) is factorized. If factorization is carried out according to expansion (6.2.2), then

$$E - \tau A - \Phi = (E - \tau A_1) \ldots (E - \tau A_m). \qquad (6.2.9)$$

It follows that

$$\Phi = E - \tau A - (E - \tau A_1) \ldots (E - \tau A_m)$$
$$= -\tau^2 \sum_{i<j} A_i A_j + \tau^3 \sum_{i<j<k} A_i A_j A_k + \cdots + (-1)^{m-1} \tau^m A_1 \ldots A_m. \qquad (6.2.10)$$

A comparison of Eqs. (6.2.5) and (6.2.10) shows that schemes (6.2.3) and (6.2.8) are equivalent, provided the initial schemes (6.2.1) and the expansion (6.2.2) are identical.

A. A. Samarskii [63] used the two-layer s.h.a. (6.1.12) as the initial scheme (6.1.1). In this case the factorized scheme is

$$(E - \alpha \tau \Lambda_1)(E - \alpha \tau \Lambda_2) \ldots (E - \alpha \tau \Lambda_m)\, u^{n+1} = \Omega\, u^n, \qquad (6.2.11)$$

where

$$\left.\begin{aligned} \Omega &= E + \tau\left[(1 - \alpha)\Lambda + \frac{h^2}{6}\sum_{i<j} \Lambda_i \Lambda_j\right] - \Phi; \\ \Phi &= -\alpha^2 \tau^2 \sum_{i<j} \Lambda_i \Lambda_j + \alpha^3 \tau^3 \sum_{i<j<k} \Lambda_i \Lambda_j \Lambda_k + \cdots \\ &\quad \cdots + (-1)^m \alpha^m \tau^m \Lambda_1 \ldots \Lambda_m. \end{aligned}\right\} \qquad (6.2.12)$$

In a particular case when $m = 2$ the Samarskii scheme takes the simple form

$$(E - \alpha \tau \Lambda_1)(E - \alpha \tau \Lambda_2)\, u^{n+1}$$
$$= [E + (1 - \alpha)\tau \Lambda_1][E + (1 - \alpha)\tau \Lambda_2]\, u^n. \qquad (6.2.13)$$

The s.h.a. for Eq. (6.1.2) in which $a_{11} = a_{22} = 1$ is derived analogously. The initial uniform scheme is

$$\frac{u^{n+1} - u^{n-1}}{2\tau} = \frac{\Lambda_{11} + \Lambda_{22}}{2}(u^{n+1} + u^{n-1}) + 2a_{12}\Lambda_{12}\, u^n -$$
$$- \frac{h^2}{12}\,\frac{u^{n+1} - 2u^n + u^{n-1}}{\tau^2} + \frac{h^2}{6}\, b\, \Lambda_1 \Lambda_2\, u^n, \qquad (6.2.14)$$

where

$$\left.\begin{aligned} \Lambda_{ii} &= \frac{\Delta_i \Delta_{-i}}{h^2}, \\ \Lambda_{12} &= \frac{\Delta_{-1}\Delta_{-2} + \Delta_1\Delta_2}{2h^2}, \quad a_{12} > 0, \\ \Lambda_{12} &= \frac{\Delta_{-1}\Delta_2 + \Delta_1\Delta_{-2}}{2h^2}, \quad a_{12} < 0, \\ b &= 1 + 2a_{12}^2 - 3\,|a_{12}|. \end{aligned}\right\} \tag{6.2.15}$$

The factorized scheme is

$$\left.\begin{aligned} A_1 A_2 \frac{u^{n+1} - u^n}{\tau} &= (1 - 2\alpha)\frac{u^n - u^{n-1}}{\tau} + \alpha\,\Lambda(u^n + u^{n-1}) + \\ &\quad + 4\alpha\, a_{12}\,\Lambda_{12}\, u^n + 2(1 - \alpha)\,\tau\, b\,\Lambda_1\,\Lambda_2\, u^n, \\ A_i = E - \alpha\tau\,\Lambda_{ii}; \quad \alpha &= \frac{1}{1 + \frac{h^2}{6\tau}}. \end{aligned}\right\} \tag{6.2.16}$$

6.3 Solution of Dirichlet's problem with the use of the schemes of higher accuracy

The schemes considered above represent integration schemes of Eqs. (6.1.1) and (6.1.2) of higher order of accuracy and they have the properties of complete consistency and absolute stability. Therefore they are fully acceptable as iterative schemes of higher accuracy.

The accuracy of integration with respect to "relaxation time" t for iterative schemes is of secondary importance, while the accuracy of approximation with respect to the space variables is essential. Therefore in the case of iterative schemes it is possible to use schemes of lower accuracy with respect to time, provided they satisfy the condition of a complete consistency and are sufficiently accurate with respect to space variables.

Consider, for simplicity, the case when $m = 2$. For Laplace's equation

$$L\,u = \frac{\partial^2 u}{\partial x_1^2} + \frac{\partial^2 u}{\partial x_2^2} = 0, \tag{6.3.1}$$

the uniform s.h.a. is

$$\Omega\,u = \left(\Lambda + \frac{h^2}{6}\Lambda_1\Lambda_2\right)u = 0. \tag{6.3.2}$$

Following A. A. Samarskii and V. B. Andreev [64] we consider a uniform absolutely stable scheme of the first order of accuracy with respect to t

$$\frac{u^{n+1} - u^n}{\tau} = \Lambda\,u^{n+1} + \frac{h^2}{6}\Lambda_1\,\Lambda_2\,u^n. \tag{6.3.3}$$

Scheme (6.3.3) is identical to

$$\frac{u^{n+1}-u^n}{\tau} = \Lambda u^n + \tau \Lambda \frac{u^{n+1}-u^n}{\tau} + \frac{h^2}{6} \Lambda_1 \Lambda_2 u^n$$

$$= \tau \Lambda \frac{u^{n+1}-u^n}{\tau} + \Omega u^n, \tag{6.3.4}$$

or to

$$(E - \tau \Lambda) \frac{u^{n+1}-u^n}{\tau} = \Omega u^n. \tag{6.3.5}$$

By factorizing the operator $E - \tau \Lambda$ we obtain a factorized s.h.a.

$$(E - \tau \Lambda_1)(E - \tau \Lambda_2) \frac{u^{n+1}-u^n}{\tau} = \Omega u^n, \tag{6.3.6}$$

which possesses a complete approximation and is strongly stable.

V. A. Enal'skii [67] obtained simple iterative s.h.a. based on the splitting schemes. We reproduce his results below in a general form.

Consider the splitting scheme

$$\left.\begin{aligned} \frac{u^{n+1/2}-u^n}{\tau} &= \Lambda_1(\alpha_1 u^{n+1/2} + \beta_1 u^n); \\ \frac{u^{n+1}-u^{n+1/2}}{\tau} &= \Lambda_2(\alpha_2 u^{n+1} + \beta_2 u^{n+1/2}) \end{aligned}\right. \tag{6.3.7}$$

with undetermined parameters α_i, β_i, $i = 1, 2$. The scheme in whole steps is

$$E - \alpha_1 \tau \Lambda_1)(E - \alpha_2 \tau \Lambda_2) u^{n+1} = (E + \beta_1 \tau \Lambda_1)(E + \beta_2 \tau \Lambda_2) u^n. \tag{6.3.8}$$

Quantities α_i, β_i are selected in such a manner that the condition of complete consistency is satisfied. Transform Eq. (6.3.8) into

$$\left.\begin{aligned} \frac{u^{n+1}-u^n}{\tau} &= \Omega_1 u^{n+1} + \Omega_2 u^n; \\ \Omega_1 &= \alpha_1 \Lambda_1 + \alpha_2 \Lambda_2 - \alpha_1 \alpha_2 \tau \Lambda_1 \Lambda_2; \\ \Omega_2 &= \beta_1 \Lambda_1 + \beta_2 \Lambda_2 + \beta_1 \beta_2 \tau \Lambda_1 \Lambda_2. \end{aligned}\right\} \tag{6.3.9}$$

The condition of a complete consistency gives

$$\Omega_1 + \Omega_2 = k \Omega = k\left(\Lambda_1 + \Lambda_2 + \frac{h^2}{6} \Lambda_1 \Lambda_2\right). \tag{6.3.10}$$

We can assume, since the parameter τ is undetermined, that $k = 1$; hen condition (6.3.10) gives

$$\begin{aligned} \alpha_1 + \beta_1 &= \alpha_2 + \beta_2 = 1; \\ \beta_1 \beta_2 - \alpha_1 \alpha_2 &= \frac{1}{6r} = \theta. \end{aligned} \tag{6.3.11}$$

From Eq. (6.3.11) we find that

$$\alpha_1 + \alpha_2 = 1 - \theta. \tag{6.3.12}$$

When

$$\alpha_1 = \alpha_2 = \alpha = \frac{1-\theta}{2} \tag{6.3.13}$$

the scheme of Samarskii (6.2.13) is obtained.

V. A. Enal'skii [68] suggested carrying out the expansion in schemes of fractional steps with operators Λ_1, Λ_2, $\Lambda_1 \Lambda_2$. Then the following scheme is obtained

$$\begin{aligned} \frac{u^{n+1/2} - u^n}{\tau} &= \Lambda_1 u^{n+1/2} + \alpha \Lambda_1 u^n + \beta \Lambda_2 u^n + \gamma \tau \Lambda_1 \Lambda_2 u^n; \\ \frac{u^{n+1} - u^{n+1/2}}{\tau} &= \Lambda_2 u^{n+1} + \alpha \Lambda_2 u^{n+1/2} + \beta \Lambda_1 u^{n+1/2} + \gamma \tau \Lambda_1 \Lambda_2 u^{n+1/2} \end{aligned} \tag{6.3.14}$$

with undetermined coefficients α, β, γ. The scheme in whole steps is

$$\left.\begin{aligned} A u^{n+1} &= B u^n; \\ A &= (E - \tau \Lambda_1)(E - \tau \Lambda_2); \\ B &= (E + \alpha \tau \Lambda_1 + \beta \tau \Lambda_2 + \gamma \tau^2 \Lambda_1 \Lambda_2) \times \\ &\quad \times (E + \beta \tau \Lambda_1 + \alpha \tau \Lambda_2 + \gamma \tau^2 \Lambda_1 \Lambda_2). \end{aligned}\right\} \tag{6.3.15}$$

The coefficients α, β, γ are selected in such a way that the operator $B - A$ is divisible by the operator $\Omega = \Lambda_1 + \Lambda_2 + (h^2/6) \Lambda_1 \Lambda_2$. This gives the following conditions for α, β, γ

$$\frac{2\gamma + (\alpha - \beta)^2 - 1}{\alpha + \beta + 1} = \sigma, \qquad \frac{\gamma^2}{\gamma(\alpha + \beta) - \alpha \beta \sigma} = \sigma, \qquad \sigma = \frac{h^2}{6\tau} = \frac{1}{6r}. \tag{6.3.16}$$

If

$$\alpha = \sigma; \qquad \beta = 1 + \sigma; \qquad \gamma = \alpha \beta, \tag{6.3.17}$$

then conditions (6.3.16) are satisfied and the scheme obtained is that of V. A. Enal'skii [68]. The general algorithm for constructing the splitting schemes of higher accuracy is given in [102]. Some a.d. schemes of higher accuracy were considered in [103, 104].

Chapter 7

Integro-differential, Integral, and Algebraic Equations

7.1 Equations of kinetics

For the kinetic theory equation (constant velocity, isotropic scattering)

$$\frac{\partial \varphi}{\partial t} + \sum_{k=1}^{m-1} u_k \frac{\partial \varphi}{\partial x_k} + \sigma \varphi = \frac{\sigma_s}{4\pi} \int \varphi(x, u, t)\, du + S(x, u, t) \quad (7.1.1)$$

the following scheme was mentioned in the work of G. I. Marchuk and the author [69] (incomplete splitting)

$$\frac{\varphi^{n+1/2} - \varphi^n}{\tau} = \Lambda_1(\alpha \varphi^{n+1/2} + \beta \varphi^n) + \bar{S}, \quad (7.1.2a)$$

$$\frac{\varphi^{n+1} - \varphi^{n+1/2}}{\tau} = \Lambda_2(\alpha \varphi^{n+1} + \beta \varphi^{n+1/2}). \quad (7.1.2b)$$

where Λ_1, Λ_2, $\bar{S}$ are approximations of the operators $\sigma E + \frac{\sigma_s}{4\pi}\int du$, $\sum_{k=1}^{m-1} u_k \frac{\partial}{\partial x_k}$, S, respectively ($\alpha \geq 0$, $\beta \geq 0$, $\alpha + \beta = 1$).

Scheme (7.1.2) is realized in the following manner. Summation of Eq. (7.1.2a) with respect to u_k results in

$$\frac{\varphi_0^{n+1/2} - \varphi_0^n}{\tau} = -\sigma_c(\alpha \varphi_0^{n+1/2} + \beta \varphi_0^n) + \bar{S}_0, \quad (7.1.3)$$

where

$$\varphi_0 = \sum \varphi \Delta u; \quad \bar{S}_0 = \sum \bar{S} \Delta u; \quad \sigma_c = \sigma - \sigma_s \quad (7.1.3a)$$

and the sum in Eq. (7.1.3a) is taken with respect to network indices.

From Eq. (7.1.3) $\varphi_0^{n+1/2}$ is determined explicitly

$$\varphi_0^{n+1/2} = \frac{1 - \beta\tau\sigma_c}{1 + \alpha\tau\sigma_c} \varphi_0^n + \frac{\bar{S}_0}{1 + \alpha\tau\sigma_c} \tau. \quad (7.1.4)$$

After this step Eq. (7.1.2a) is integrated to give

$$\frac{\varphi^{n+1/2} - \varphi^n}{\tau} + \sigma(\alpha \varphi^{n+1/2} + \beta \varphi^n) = \frac{\sigma_s}{4\pi}(\alpha \varphi_0^{n+1/2} + \beta \varphi_0^n) + \bar{S}. \quad (7.1.2c)$$

In order to simplify the execution of the second fractional step a scheme of complete splitting is used

$$\frac{\varphi^{n+1/m} - \varphi^n}{\tau} = \Lambda_1(\alpha \varphi^{n+1/m} + \beta \varphi^n) + \bar{S}, \quad (7.1.5a)$$

$$\frac{\varphi^{n+(i+1)/m} - \varphi^{n+i/m}}{\tau} = \Lambda_{2i}(\alpha \varphi^{n+(i+1)/m} + \beta \varphi^{n+i/m}), \quad i = 1, \ldots, m-1, \quad (7.1.5b)$$

where $\Lambda_2 = \Lambda_{21} + \cdots + \Lambda_{2\,m-1}$; $\Lambda_{2\,i}$ is the approximation of the one-dimensional operators $u_i(\partial/\partial x_i)$, $i = 1, \ldots, m-1$; $(m-1)$ is the space dimension.

The boundary conditions should now be analyzed. A mixed Cauchy problem is used for Eq. (7.1.1)

$$\begin{aligned} &\varphi(x, u, 0) = \varphi_0(x, u), \qquad (x, u) \in G, \\ &\varphi(x, u, t) = 0, \end{aligned} \tag{7.1.6}$$

$$\sum_{k=1}^{m-1} u_k n_k \leq 0, \qquad (x, u, t) \in \Gamma \tag{7.1.7}$$

in the cylindrical region $\Pi = G \times H$ with the base G, side boundary $\Gamma = \gamma \times H$, $\gamma = \bar{G} - G$ of normal $n = \{n_k\}$. If Π is a parallelepiped, then the satisfaction of boundary conditions in the scheme of complete splitting is obvious, namely, at each i-th fractional step of Eq. (7.1.5 b), a scheme of running calculation along x_i in the direction from the illuminated edge $\left(\sum_{k=1}^{m-1} u_k n_k \leq 0\right)$ is carried out.

In the case of an arbitrary cylindrical region Π_0 with base G_0, the region Π_0 is included in the parallelepiped Π and the value of φ is defined respectively in the region $\Pi - \Pi_0$. In the two-dimensional case the definition of φ is carried out in the following way (Fig. 7). At the first fractional step ($i = 1$) of Eq. (7.1.5 b) it is assumed that $\varphi^{n+2/3} = 0$ at AB and $\varphi^{n+1/3} = 0$ in regions FBK and KAE. The second fractional step ($i = 2$) is performed with the initial data $\varphi^{n+2/3}$ which were determined from the first fractional step and with boundary conditions $\varphi = \varphi^{n+1}$ at AD. A detailed study of the convergence of the different scheme is given in [70, 71].

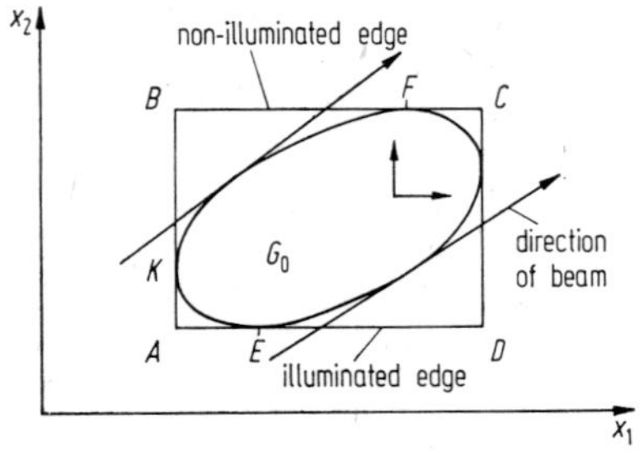

Fig. 7. Extension of the definition of φ in $\Pi - \Pi_0$ in the two-dimensional case. The small arrows show the direction from the illuminated to the non-illuminated edge

An analogous approach to the summation of φ, leading to a difference equation which contains only φ_0 in the upper layer was used earlier by V. Ya. Gol'din [98].

7.2 Algebraic equations

A. A. Samarskii [50] suggested an algorithm for the solution of a system of differential equations

$$\frac{\partial u_i}{\partial t} + \sum_{j=1}^{m} a_{ij} u_j = f_i(t), \quad i, j = 1, \ldots, m. \tag{7.2.1}$$

An explicit integration scheme of Eq. (7.2.1) is of first order accuracy and requires $O(m^2)$ operations.

The ordinary implicit scheme

$$\frac{u^{n+1} - u^n}{\tau} + A \frac{u^{n+1} + u^n}{2} = f^{n+1}, \quad u = \{u_i\}, \quad f = \{f_i\}, \quad A = \{a_{ij}\} \tag{7.2.2}$$

is of second order accuracy and requires $O(m^3)$ operations. A scheme which is analogous to the a.d. scheme was suggested in [50]

$$\begin{aligned} \frac{u^{n+1/2} - u^n}{\tau/2} + A_1 u^{n+1/2} + A_2 u^n &= f^{n+1/2}, \\ \frac{u^{n+1} - u^{n+1/2}}{\tau/2} + A_1 u^{n+1/2} + A_2 u^{n+1} &= f^{n+1/2}, \end{aligned} \tag{7.2.3}$$

where

$$\left.\begin{aligned} &A = A_1 + A_2, \\ A_1 = (a_{ij}^-), \quad &A_2 = (a_{ij}^+), \quad a_{ij}^- = 0, \quad j > i, \\ &a_{ii}^- + a_{ii}^+ = a_{ii}. \end{aligned}\right\} \tag{7.2.4}$$

If the triangular matrices A_1 and A_2 are positive definite, i.e.,

$$(A_\alpha u, u) \geq C \|u\|^2, \quad C > 0, \quad \alpha = 1, 2, \tag{7.2.5}$$

then scheme (7.2.3) is stable.

Scheme (7.2.3) can also be considered as iterative. For the coefficient of error decay ϱ (see [50]) we have

$$\varrho \leq \bar{\varrho} = \frac{1 - \sigma}{1 + \sigma}, \quad \sigma = \frac{2C\tau}{1 + d^2 \tau^2}, \tag{7.2.6}$$

where

$$d = \max(\|A_1\|, \|A_2\|).$$

It is easy to see that a scheme analogous to the splitting scheme can also be used here.

Chapter 8

Some Problems of Hydrodynamics

The method of fractional steps can be successfully applied to problems of hydrodynamics for the construction of economical integration and iterative schemes.

8.1 Potential flow past a contour

Consider the plane potential flow of an incompressible liquid past a contour γ with an axis of symmetry $x_2 = 0$ (Fig. 8). The flow at

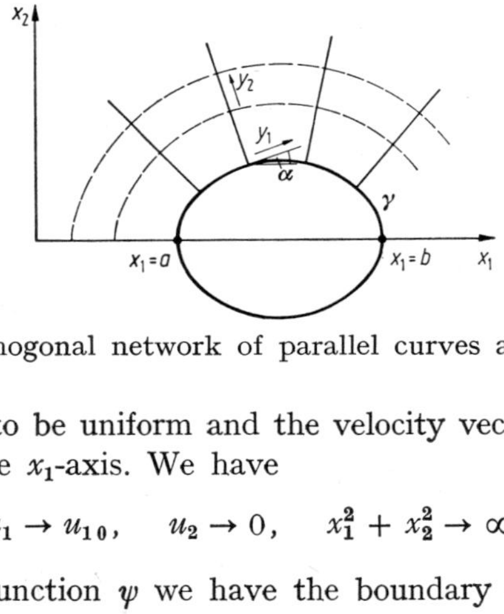

Fig. 8. Orthogonal network of parallel curves and normals

infinity is taken to be uniform and the velocity vector $\boldsymbol{u} = \{u_1, u_2\}$ is directed along the x_1-axis. We have

$$u_1 \to u_{10}, \quad u_2 \to 0, \quad x_1^2 + x_2^2 \to \infty. \tag{8.1.1}$$

For the stream function ψ we have the boundary value problem

$$\Delta\psi = \frac{\partial^2\psi}{\partial x_1^2} + \frac{\partial^2\psi}{\partial x_2^2} = 0, \tag{8.1.2}$$

$$\psi(x_1, 0) = 0, \quad -\infty < x_1 \leq a, \quad b \leq x_1 < \infty, \tag{8.1.3a}$$

$$\psi(x_1, x_2) = 0, \quad (x_1, x_2) \in \gamma, \tag{8.1.3b}$$

$$\frac{\partial\psi}{\partial x_2} = -u_{10}, \quad \frac{\partial\psi}{\partial x_1} = 0, \quad x_1^2 + x_2^2 \to \infty. \tag{8.1.3c}$$

The problem (8.1.2), (8.1.3) can be solved in Cartesian coordinates but the network in the vicinity of the boundary will be non-uniform in this case and this reduces the accuracy of the calculations.

Let us introduce an orthogonal system of coordinates y_1, y_2, where y_1 is the length of the contour arc γ; y_2 is the distance from γ measured along the normal; y_1 and y_2 are related to x_1 and x_2 by the equations

$$x_1 = x_1(y_1) - y_2 \sin\alpha(y_1),$$

$$x_2 = x_2(y_1) + y_2 \cos\alpha(y_1), \qquad \cos\alpha = \frac{dx_1}{dy_1},$$

$$dx_1^2 + dx_2^2 = [1 - y_2 k]^2 dy_1^2 + dy_2^2,$$

where

$$x_1 = x_1(y_1), \qquad x_2 = x_2(y_1)$$

is the parametric representation of the contour γ; $k(y_1) = d\alpha/dy_1$ is the contour curvature.

Eq. (8.1.2) is transformed into

$$\frac{\partial}{\partial y_1}\left(\frac{H_2}{H_1}\frac{\partial\psi}{\partial y_1}\right) + \frac{\partial}{\partial y_2}\left(\frac{H_1}{H_2}\frac{\partial\psi}{\partial y_2}\right) = 0, \tag{8.1.4}$$

where

$$H_1 = (1 - k y_2)^2, \qquad H_2 = 1. \tag{8.1.5}$$

The integration region is transformed into a semi-infinite strip

$$0 \le y_1 \le l, \qquad 0 \le y_2 < \infty, \tag{8.1.6}$$

where l is the half perimeter of the contour.

The boundary conditions are given in the y_1, y_2 plane, as follows:

$$\psi(0, y_2) = 0, \qquad \psi(l, y_2) = 0, \tag{8.1.7a}$$

$$\psi(y_1, 0) = 0, \qquad 0 \le y_1 \le l, \tag{8.1.7b}$$

$$\psi(y_1, a) = -u_{10} x_2 = -u_{10}\{x_2(y_1) + a\cos\alpha(y_1)\}. \tag{8.1.7c}$$

An exact representation of the boundary conditions is obtained at $a \to \infty$ if condition (8.1.7c) is replaced by Eq. (8.1.3c) and if Eq. (8.1.3c) is expressed in y_1 and y_2 coordinates.

The Dirichlet problem (8.1.4), (8.1.7) in a rectangle $0 \le y_1 \le l$, $0 \le y_2 \le a$ is solved by a known method (see Ch. 4).

In order to speed up the convergence the scale along the y_2-axis should be changed by the introduction of a new variable $Y_2 = f(y_2)$. Furthermore, it is convenient to introduce the relaxation factor $F(y_1, y_2, t)$ in such a way as to solve the Dirichlet problem for the unsteady equation

$$\frac{\partial\psi}{\partial t} = F(y_1, y_2, t)\left[\frac{\partial}{\partial y_1}\frac{H_2}{H_1}\frac{\partial\psi}{\partial y_1} + \frac{\partial}{\partial y_2}\frac{H_1}{H_2}\frac{\partial\psi}{\partial y_2}\right]. \tag{8.1.8}$$

An algorithm of similar type was worked out in the form of a program which also allows for the possibility of axial symmetry.

8.2 Potential flow of an incompressible heavy liquid with a free boundary (spillway problem)

Consider the problem of flow of a stream, uniform at infinity, over an irregular bottom (Fig. 9).

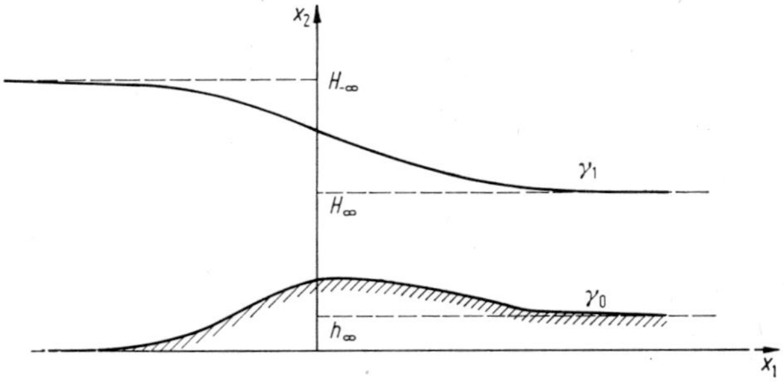

Fig. 9. Flow of a stream over an irregular bottom

We suppose that the bottom surface is represented by the equation

$$x_2 = f(x_1), \tag{8.2.1}$$

where $f(x_1)$ satisfies the conditions

$$f(-\infty) = 0; \quad f(+\infty) = h_\infty. \tag{8.2.2}$$

The motion of the liquid satisfies, as before, Eq. (8.1.2).

The boundary conditions in the physical plane x_1, x_2 are

$$u_1 = -\frac{\partial \psi}{\partial x_2} = u_{-\infty}, \quad u_2 = 0, \quad 0 \le x_2 \le H_{-\infty}, \quad x_1 = -\infty, \tag{8.2.3a}$$

$$u_1 = -\frac{\partial \psi}{\partial x_2} = u_\infty, \quad u_2 = 0, \quad h_\infty \le x_2 \le H_\infty, \quad x_1 = +\infty, \tag{8.2.3b}$$

$$\psi = 0, \quad (x_1, x_2) \in \gamma_0,$$

$$\frac{1}{2}\left[\left(\frac{\partial \psi}{\partial x_1}\right)^2 + \left(\frac{\partial \psi}{\partial x_2}\right)^2\right] + P_0 = -g x_2 + c_1, \quad (x_1, x_2) \in \gamma_1, \tag{8.2.3c}$$

$$\psi = \text{const} = c_2, \quad (x_1, x_2) \in \gamma_1. \tag{8.2.3d}$$

Here $H_{-\infty}$ is a given quantity; P_0 is the pressure at the upper (free) boundary.

Condition (8.2.3d) means that the free surface in steady flow is, like the bottom, a streamline (ψ-line); and condition (8.2.3c) corresponds to Bernoulli's equation.

The free surface is undetermined and should be selected in such a way that conditions (8.2.3c) and (8.2.3d) are compatible. Constants c_1 and c_2 are found from the conditions

$$c_2 = u_{-\infty} H_{-\infty}, \tag{8.2.4}$$

$$c_1 = \frac{1}{2} u_{-\infty}^2 + P_0 + g H_{-\infty}. \tag{8.2.5}$$

Finally, the ultimate height H_∞ and velocity u_∞ are determined from the laws of conservation of mass and energy (Bernoulli's law)

$$u_{-\infty} H_\infty = u_\infty (H_\infty - h_\infty), \tag{8.2.6}$$

$$c_1 - P_0 = \frac{1}{2} u_{+\infty}^2 + g H_\infty. \tag{8.2.7}$$

New independent variables x_1, ψ are needed for the solution of the complex non-linear problem (8.1.2), (8.2.3) and x_2 is selected as the unknown function. Instead of Eq. (8.1.2) a quasilinear elliptic equation is obtained.

For convenience we use the notation: $x_1 = x$, $x_2 = z$. Then the equation for z is

$$\frac{\partial^2 z}{\partial x^2} - \frac{\partial}{\partial \psi} \frac{1 + \left(\frac{\partial z}{\partial x}\right)^2}{\frac{\partial z}{\partial \psi}} = 0 \tag{8.2.8}$$

or

$$L z = z_\psi^2 \frac{\partial^2 z}{\partial x^2} - 2 z_x z_\psi \frac{\partial^2 z}{\partial x \partial \psi} + [1 + z_x^2] \frac{\partial^2 z}{\partial \psi^2} = 0. \tag{8.2.9}$$

The region of integration in the variables x, ψ becomes rectangular. Boundary conditions (8.2.3) are in this case (Fig. 10)

$$z = -\frac{\psi}{u_{-\infty}}, \quad 0 \le \psi \le c_2, \quad x = -c, \tag{8.2.10a}$$

$$z = \frac{\psi}{u_\infty} + h_\infty, \quad 0 \le \psi \le c_2, \quad x = +c, \tag{8.2.10b}$$

$$z = f(x), \quad -c \le x \le c, \quad \psi = 0, \tag{8.2.10c}$$

$$z_x^2 + (g z - c_1 + P_0) z_\psi^2 + 1 = 0, \quad -c \le x \le c, \quad \psi = c_2. \tag{8.2.10d}$$

An exact form of this problem is obtained for $c \to \infty$.

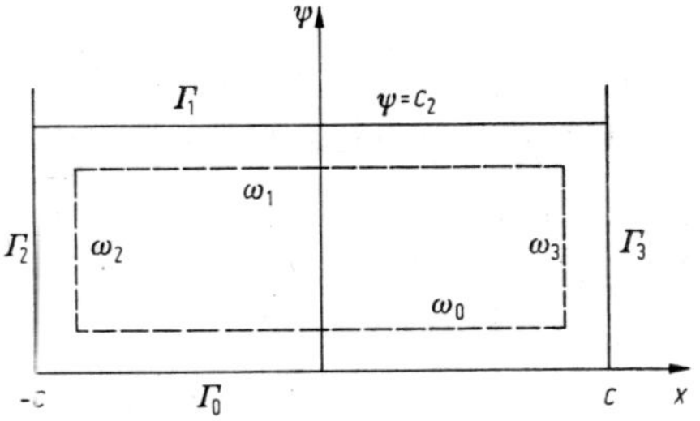

Fig. 10. Integration region in the plane x, ψ

Values of z on the straight lines Γ_0, Γ_2, Γ_3 are given by Eqs. (8.2.10a, b, c) and the nonlinear condition (8.2.10d) is given along the straight line Γ_1. A complex iterative process is needed for the solution of the resulting nonlinear problem. For this, as usual, the following unsteady equation is considered, corresponding to Eq. (8.2.9)

$$\frac{\partial z}{\partial t} = L z. \tag{8.2.11}$$

This equation is integrated with some scheme in fractional steps. The coefficients in the L operator are taken in this case from the preceding iteration. For example, the iterative splitting scheme is

$$\left.\begin{aligned} \frac{z^{n+1/2} - z^n}{\tau} &= (z_\psi^2)^n \frac{\Delta_1 \Delta_{-1}}{h_1^2} z^{n+1/2} - (z_x z_\psi)^n \frac{\Delta_1 + \Delta_{-1}}{2h_1} \frac{\Delta_2 + \Delta_{-2}}{2h_2} z^n; \\ \frac{z^{n+1} - z^{n+1/2}}{\tau} &= -(z_x z_\psi)^{n+1/2} \frac{\Delta_1 + \Delta_{-1}}{2h_1} \frac{\Delta_2 + \Delta_{-2}}{2h_2} z^{n+1/2} + \\ &\quad + [1 + z_x^2]^{n+1/2} \frac{\Delta_2 \Delta_{-2}}{h_2^2} z^{n+1}. \end{aligned}\right\} \tag{8.2.12}$$

The boundary condition (8.2.10d) is linearized in a similar way

$$(z_x)^k z_x^{n+1} + [(g z - c_1 + P_0) z_\psi]^k z_\psi^{n+1} + 1 = 0, \tag{8.2.13}$$

where k is the iteration index at the boundary.

The network lines nearest to the boundary $\Gamma_0, \Gamma_1, \Gamma_2, \Gamma_3$, are denoted by $\omega_0, \omega_1, \omega_2, \omega_3$.

Let the value of z, from the n-th iteration, be given on Γ_1. Then for z a boundary value problem of the first type is obtained, and, using the given boundary conditions, one or several cycles are performed according to scheme (8.2.12). After this the conditions on Γ_1 are recalculated by iterations according to the scheme

$$\frac{z^{k+1} - z^k}{\tau} = (z_x)^k \frac{\Delta_1 + \Delta_{-1}}{2h_1} z^{k+1} + [(g z - c_1 + P_0) z_\psi]^k \frac{z^{k+1} - z^n(\omega_1)}{h_2} + 1, \tag{8.2.14}$$

where $z^n(\omega_1)$ is taken on ω_1 from the last iteration in the solution of the Dirichlet problem. By alternating the iteration cycles a convergent iterative process is obtained. This algorithm was suggested and realized in [73].

8.3 The flow of a viscous liquid

In the two-dimensional case the flow of a compressible viscous liquid is described by the equations

$$\varrho\left(\frac{\partial u_1}{\partial t} + u_1 \frac{\partial u_1}{\partial x_1} + u_2 \frac{\partial u_1}{\partial x_2}\right) + \frac{\partial p}{\partial x_1} = \mu \Delta u_1 + \left(\zeta + \frac{\mu}{3}\right) \frac{\partial}{\partial x_1}\left(\frac{\partial u_1}{\partial x_1} + \frac{\partial u_2}{\partial x_2}\right); \tag{8.3.1 a}$$

$$\varrho\left(\frac{\partial u_2}{\partial t} + u_1 \frac{\partial u_2}{\partial x_1} + u_2 \frac{\partial u_2}{\partial x_2}\right) + \frac{\partial p}{\partial x_2} = \mu \Delta u_2 + \left(\zeta + \frac{\mu}{3}\right) \frac{\partial}{\partial x_2}\left(\frac{\partial u_1}{\partial x_1} + \frac{\partial u_2}{\partial x_2}\right); \tag{8.3.1 b}$$

$$\frac{\partial \varrho}{\partial t} + u_1 \frac{\partial \varrho}{\partial x_1} + u_2 \frac{\partial \varrho}{\partial x_2} + \varrho\left(\frac{\partial u_1}{\partial x_1} + \frac{\partial u_2}{\partial x_2}\right) = 0, \tag{8.3.1 c}$$

where μ and ζ are the coefficients of viscosity.

For an incompressible liquid we have

$$\frac{d\varrho}{dt} = \frac{\partial \varrho}{\partial t} + u_1 \frac{\partial \varrho}{\partial x_1} + u_2 \frac{\partial \varrho}{\partial x_2} = 0, \tag{8.3.2}$$

and Eqs. (8.3.1 a, b, c) take the form

$$\varrho\left(\frac{\partial u_1}{\partial t} + u_1 \frac{\partial u_1}{\partial x_1} + u_2 \frac{\partial u_1}{\partial x_2}\right) + \frac{\partial p}{\partial x_1} = \mu \Delta u_1; \tag{8.3.3 a}$$

$$\varrho\left(\frac{\partial u_2}{\partial t} + u_1 \frac{\partial u_2}{\partial x_1} + u_2 \frac{\partial u_2}{\partial x_2}\right) + \frac{\partial p}{\partial x_2} = \mu \Delta u_2; \tag{8.3.3 b}$$

$$\frac{\partial u_1}{\partial x_1} + \frac{\partial u_2}{\partial x_2} = 0. \tag{8.3.3 c}$$

Here our calculations are confined to the special case of incompressible flow. This is described by Eqs. (8.3.3 a, b, c) or, equivalently, the equations

$$u_1 \frac{\partial u_1}{\partial x_1} + u_2 \frac{\partial u_1}{\partial x_2} + \frac{1}{\varrho} \frac{\partial p}{\partial x_1} = \nu \Delta u_1; \tag{8.3.4a}$$

$$u_1 \frac{\partial u_2}{\partial x_1} + u_2 \frac{\partial u_2}{\partial x_2} + \frac{1}{\varrho} \frac{\partial p}{\partial x_2} = \nu \Delta u_2; \tag{8.3.4b}$$

$$\frac{\partial u_1}{\partial x_1} + \frac{\partial u_2}{\partial x_2} = 0, \tag{8.3.4c}$$

where $\nu = \mu/\varrho$ is the coefficient of kinematic viscosity.

Boundary conditions in this case correspond to symmetric flow (Fig. 11).

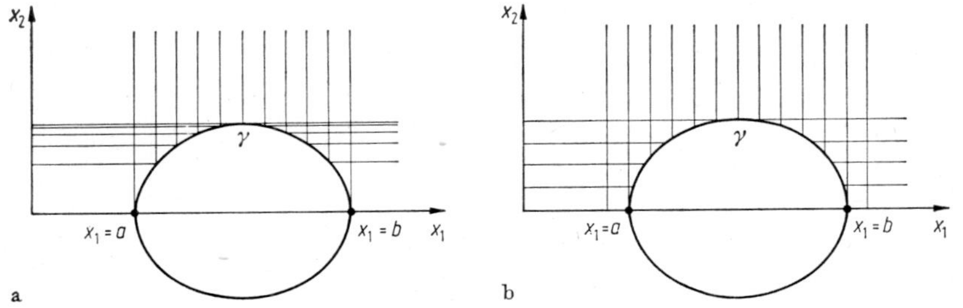

Fig. 11. a — network adapted to the contour, b — network not adapted to the contour

To construct a relaxation process some unsteady problem is considered in accordance with Eqs. (8.3.4a, b, c).

A model of a weakly compressible liquid can serve as the most natural relaxation model. In this case we start with system (8.3.1). Retaining Eqs. (8.3.1 a) and (8.3.1 b) we assume that

$$p = a^2 \varrho^k \tag{8.3.5}$$

(a^2, k are constants). Eq. (8.3.1b) can be written in the form

$$\varepsilon\left(\frac{\partial p}{\partial t}+u_1\frac{\partial p}{\partial x_1}+u_2\frac{\partial p}{\partial x_2}\right)+p\left(\frac{\partial u_1}{\partial x_1}+\frac{\partial u_2}{\partial x_2}\right)=0, \qquad \varepsilon=\frac{1}{k}. \tag{8.3.6}$$

When k is sufficiently large an equation with a small parameter ε is obtained.

The unsteady system in this case is

$$\left.\begin{aligned}&\frac{\partial u_1}{\partial t}+u_1\frac{\partial u_1}{\partial x_1}+u_2\frac{\partial u_1}{\partial x_2}+\frac{1}{\varrho}\frac{\partial p}{\partial x_1}=\nu\Delta u_1;\\&\frac{\partial u_2}{\partial t}+u_1\frac{\partial u_2}{\partial x_1}+u_2\frac{\partial u_2}{\partial x_2}+\frac{1}{\varrho}\frac{\partial p}{\partial x_2}=\nu\Delta u_2;\\&\varepsilon\left(\frac{\partial p}{\partial t}+u_1\frac{\partial p}{\partial x_1}+u_2\frac{\partial p}{\partial x_2}\right)+p\left(\frac{\partial u_1}{\partial x_1}+\frac{\partial u_2}{\partial x_2}\right)=0.\end{aligned}\right\} \tag{8.3.7}$$

It is assumed below that $\varrho = 1$.

A splitting scheme in coordinates x_1 and x_2 can be applied to Eq. (8.3.7), approximating the system

$$\left.\begin{aligned}&\frac{1}{2}\frac{\partial u_1}{\partial t}+u_1\frac{\partial u_1}{\partial x_1}+\frac{\partial p}{\partial x_1}=\nu\frac{\partial^2 u_1}{\partial x_1^2};\\&\frac{1}{2}\frac{\partial u_2}{\partial t}+u_1\frac{\partial u_2}{\partial x_1}=\nu\frac{\partial^2 u_2}{\partial x_1^2};\\&\frac{1}{2}\varepsilon\frac{\partial p}{\partial t}+\varepsilon u_1\frac{\partial p}{\partial x_1}+p\frac{\partial u_1}{\partial x_1}=0\end{aligned}\right\} \tag{8.3.8}$$

at the first half-step

$$t\in[\tau,(n+1/2)\tau],$$

and the system

$$\left.\begin{aligned}&\frac{1}{2}\frac{\partial u_1}{\partial t}+u_2\frac{\partial u_1}{\partial x_2}=\nu\frac{\partial^2 u_1}{\partial x_2^2};\\&\frac{1}{2}\frac{\partial u_2}{\partial t}+u_2\frac{\partial u_2}{\partial x_2}+\frac{\partial p}{\partial x_2}=\nu\frac{\partial^2 u_2}{\partial x_2^2};\\&\frac{1}{2}\varepsilon\frac{\partial p}{\partial t}+\varepsilon u_2\frac{\partial p}{\partial x_2}+p\frac{\partial u_2}{\partial x_2}=0\end{aligned}\right\} \tag{8.3.9}$$

at the second half-step

$$t\in[(n+1/2)\tau,(n+1)\tau].$$

Still another unsteady system can be used. This is of the type

$$\begin{aligned}&\frac{\partial u_1}{\partial t}+u_1\frac{\partial u_1}{\partial x_1}+u_2\frac{\partial u_1}{\partial x_2}+\frac{\partial p}{\partial x_1}=\nu\Delta u_1;\\&\frac{\partial u_2}{\partial t}+u_1\frac{\partial u_2}{\partial x_1}+u_2\frac{\partial u_2}{\partial x_2}+\frac{\partial p}{\partial x_2}=\nu\Delta u_2;\end{aligned} \tag{8.3.10}$$

$$\frac{\partial q}{\partial t}+\frac{\partial u_1}{\partial x_1}+\frac{\partial u_2}{\partial x_2}=0, \tag{8.3.11}$$

where

$$q = \frac{1}{2} q_1 + \frac{1}{2} q_2; \quad q_1 = p + \frac{u_1^2}{2}; \quad q_2 = p + \frac{u_2^2}{2}. \tag{8.3.12}$$

The splitting scheme is applied to the system

$$\left.\begin{aligned} \frac{1}{2}\frac{\partial u_1}{\partial t} + \frac{\partial q_1}{\partial x_1} &= \nu \frac{\partial^2 u_1}{\partial x_1^2}; \\ \frac{1}{2}\frac{\partial u_2}{\partial t} + u_1 \frac{\partial u_2}{\partial x_1} &= \nu \frac{\partial^2 u_2}{\partial x_1^2}; \\ \frac{1}{2}\frac{\partial q_1}{\partial t} + \frac{\partial u_1}{\partial x_1} &= 0, \end{aligned}\right\} \tag{8.3.13}$$

$$\left.\begin{aligned} \frac{1}{2}\frac{\partial u_1}{\partial t} + u_2 \frac{\partial u_1}{\partial x_2} &= \nu \frac{\partial^2 u_1}{\partial x_2^2}; \\ \frac{1}{2}\frac{\partial u_2}{\partial t} + \frac{\partial q_2}{\partial x_2} &= \nu \frac{\partial^2 u_2}{\partial x_2^2}; \\ \frac{1}{2}\frac{\partial q_2}{\partial t} + \frac{\partial u_2}{\partial x_2} &= 0. \end{aligned}\right\} \tag{8.3.14}$$

The corresponding splitting scheme is

$$\left.\begin{aligned} &\frac{u_1^{n+1/2} - u_1^n}{\tau} + \frac{\Delta_1}{h_1}(\alpha q_1^{n+1/2} + \beta q_1^n) = \nu \frac{\Delta_1 \Delta_{-1}}{h_1^2}(\alpha u_1^{n+1/2} + \beta u_1^n); \\ &\frac{u_2^{n+1/2} - u_2^n}{\tau} + (\alpha u_1^n + \beta u_1^{n+1/2}) \frac{\Delta}{h_1}(\alpha u_2^{n+1/2} + \beta u_2^n) \\ &\qquad = \nu \frac{\Delta_1 \Delta_{-1}}{h_1^2}(\alpha u_2^{n+1/2} + \beta u_2^n); \\ &\frac{q_1^{n+1/2} - q_1^n}{\tau} + \frac{\Delta_{-1}}{h_1}(\alpha u_1^{n+1/2} + \beta u_1^n) = 0, \end{aligned}\right\} \tag{8.3.15}$$

$$\left.\begin{aligned} &\frac{u_1^{n+1} - u_1^{n+1/2}}{\tau} + (\alpha u_2^{n+1} + \beta u_2^{n+1/2}) \frac{\delta}{h_2}(\alpha u_1^{n+1} + \beta u_1^{n+1/2}) \\ &\qquad = \nu \frac{\Delta_2 \Delta_{-2}}{h_2^2}(\alpha u_1^{n+1} + \beta u_1^{n+1/2}); \\ &\frac{u_2^{n+1} - u_2^{n+1/2}}{\tau} + \frac{\Delta_2}{h_2}(\alpha q_2^{n+1} + \beta q_2^{n+1/2}) \\ &\qquad = \nu \frac{\Delta_2 \Delta_{-2}}{h_2^2}(\alpha u_2^{n+1} + \beta u_2^{n+1/2}); \\ &\frac{q_2^{n+1} - q_2^{n+1/2}}{\tau} + \frac{\Delta_{-2}}{h_2}(\alpha u_2^{n+1} + \beta u_2^{n+1/2}) = 0, \end{aligned}\right\} \tag{8.3.16}$$

where

$$\begin{aligned} \Delta &= \Delta_{-1}; \quad (\alpha u_1^n + \beta u_1^{n+1/2}) \geq 0; \\ \Delta &= \Delta_1; \quad (\alpha u_1^n + \beta u_1^{n+1/2}) < 0; \\ \delta &= \Delta_{-2}; \quad (\alpha u_2^{n+1} + \beta u_2^{n+1/2}) \geq 0; \\ \delta &= \Delta_2; \quad (\alpha u_2^{n+1} + \beta u_2^{n+1/2}) < 0; \\ &\alpha \geq 0; \quad \beta \geq 0; \quad \alpha + \beta = 1. \end{aligned}$$

Δ, δ can also be central differences. It is easy to see that the implicit schemes (8.3.15) and (8.3.16) can be reduced to three-point sweeps along the network lines (Fig. 11). The network can be either adapted to the contour (network lines meet at the contour) (Fig. 11a), or not adapted (Fig. 11b). In the latter case additional interpolation is used.

In practice, the condition at $x = \mp\infty$ is applied at $x = \mp c$, with c sufficiently large. Such an algorithm was studied by B. G. Kuznetsov and the author [74], and realized in program form by N. N. Vladimirova. New splitting schemes with artificial relaxation operator were considered in [107, 108].

8.4 The method of channel flows

A curvilinear network is used in the solution of many problems of hydrodynamics. The method of channel flows consists of a simultaneous transformation of coordinates and velocity components in such a way that each component characterizes the flow velocity in the channel of corresponding coordinates (Fig. 12).

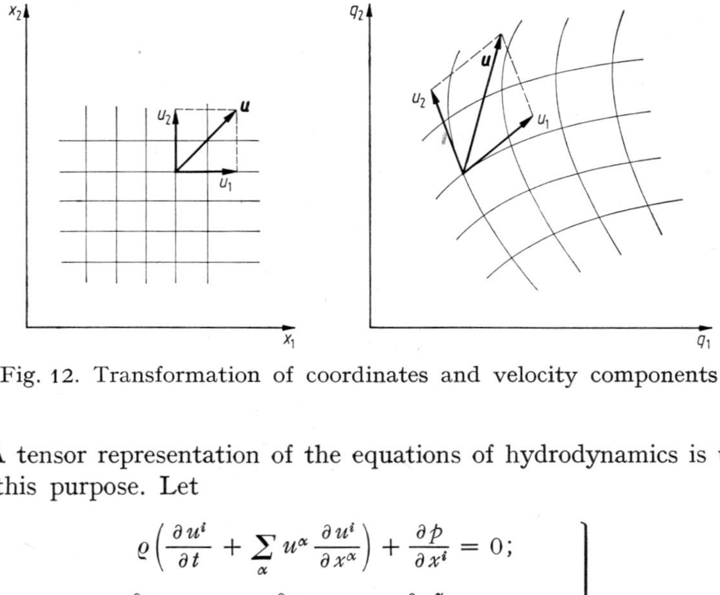

Fig. 12. Transformation of coordinates and velocity components

A tensor representation of the equations of hydrodynamics is used for this purpose. Let

$$\left.\begin{aligned} &\varrho\left(\frac{\partial u^i}{\partial t} + \sum_\alpha u^\alpha \frac{\partial u^i}{\partial x^\alpha}\right) + \frac{\partial p}{\partial x^i} = 0; \\ &\frac{\partial \varrho}{\partial t} + \sum_\alpha u^\alpha \frac{\partial \varrho}{\partial x^\alpha} + \varrho \sum_\alpha \frac{\partial u^\alpha}{\partial x^\alpha} = 0; \\ &\frac{\partial S}{\partial t} + \sum_\alpha u^\alpha \frac{\partial S}{\partial x^\alpha} = 0, \qquad p = p(\varrho, S) \end{aligned}\right\} \tag{8.4.1}$$

be a system of equations of gas dynamics in orthogonal Cartesian coordinates. We consider u^i as the contravariant components of the velocity vector $\boldsymbol{u}$, and $\partial p/\partial x^i$ as the contravariant gradient p.

Let the functions

$$q^i = q^i(x^1, \ldots, x^m, t), \qquad i = 1, \ldots, m, \tag{8.4.2}$$

determine the transformation from coordinates x^i to curvilinear coordinates q^i at a given time t. By introducing, as usual, the transformation matrices

$$a^i_j = \frac{\partial q^i}{\partial x^j}; \qquad A^i_j = \frac{\partial x^i}{\partial q^j}; \qquad \sum_\alpha a^i_\alpha A^\alpha_j = \delta^i_j, \tag{8.4.3}$$

the following transformation law is obtained

$$U^i = \sum_\alpha a^i_\alpha u^\alpha, \tag{8.4.4}$$

where U^i represents components of the velocity vector which correspond to coordinates q^i. It follows that

$$\frac{\partial U^i}{\partial t} = \sum_\alpha \left(a^i_\alpha \frac{\partial u^\alpha}{\partial t} + \frac{\partial a^i_\alpha}{\partial t} u^\alpha\right) = \sum_\alpha a^i_\alpha \frac{\partial u^\alpha}{\partial t} + \sum_{\alpha,\beta} \frac{\partial a^i_\beta}{\partial t} A^\beta_\alpha U^\alpha$$

$$= \sum_\alpha a^i_\alpha \frac{\partial u^\alpha}{\partial t} + \sum_\alpha b^i_\alpha U^\alpha, \qquad b^i_\alpha = \sum_\beta \frac{\partial a^i_\beta}{\partial t} A^\beta_\alpha. \tag{8.4.5}$$

Thus, the expressions

$$\frac{\partial U^i}{\partial t} - \sum_\alpha b^i_\alpha U^\alpha \tag{8.4.6}$$

are transformed as components of a contravariant vector, and after transformation to coordinates q^i according to (8.4.2) and to components U^i according to Eq. (8.4.5), Eqs. (8.4.1) become

$$\left.\begin{aligned} \varrho\left(\frac{\partial U^i}{\partial t} + \sum_\alpha U^\alpha \frac{\partial U^i}{\partial q^\alpha}\right) + \sum_\alpha g^{i\alpha} \frac{\partial p}{\partial q^\alpha} &= \Phi^i; \\ \frac{\partial \varrho}{\partial t} + \sum_\alpha U^\alpha \frac{\partial \varrho}{\partial q^\alpha} + \varrho \sum_\alpha \frac{\partial U^\alpha}{\partial q^\alpha} &= \Psi; \\ \frac{\partial S}{\partial t} + \sum_\alpha U^\alpha \frac{\partial S}{\partial q^\alpha} = 0, \qquad i, \alpha = 1, \ldots, m,& \end{aligned}\right\} \tag{8.4.7}$$

where

$$\left.\begin{aligned} \Phi^i &= -\varrho\Big(\sum_\alpha b^i_\alpha U^\alpha + \sum_{\alpha,\beta} U^\beta \Gamma^i_{\alpha\beta} U^\alpha\Big); \\ \Psi &= -\varrho \sum_{\beta,\gamma} \Gamma^\gamma_{\gamma\beta} U^\beta; \\ \Gamma^i_{\alpha\beta} &= \sum_\gamma a^i_\gamma \frac{\partial A^\gamma_\alpha}{\partial q^\beta}, \qquad i, \alpha, \beta, \gamma = 1, \ldots, m. \end{aligned}\right\} \tag{8.4.8}$$

In the particular case $m = 2$ consider a splitting scheme

$$\frac{1}{2}\varrho\frac{\partial U^1}{\partial t} + \varrho U^1\frac{\partial U^1}{\partial q^1} + g^{11}\frac{\partial p}{\partial q^1} = \frac{1}{2}\Phi^1; \tag{8.4.9a}$$

$$\frac{1}{2}\varrho\frac{\partial U^2}{\partial t} + \varrho U^1\frac{\partial U^2}{\partial q^1} + g^{21}\frac{\partial p}{\partial q^1} = \frac{1}{2}\Phi^2; \tag{8.4.9b}$$

$$\frac{1}{2}\frac{\partial \varrho}{\partial t} + U^1\frac{\partial \varrho}{\partial q^1} + \varrho\frac{\partial U^1}{\partial q^1} = \frac{1}{2}\Psi; \tag{8.4.9c}$$

$$\frac{1}{2}\frac{\partial S}{\partial t} + U^1\frac{\partial S}{\partial q^1} = 0, \tag{8.4.9d}$$

$$\left.\begin{aligned}
\frac{1}{2}\varrho\frac{\partial U^1}{\partial t} + \varrho U^2\frac{\partial U^1}{\partial q^2} + g^{12}\frac{\partial p}{\partial q^2} &= \frac{1}{2}\Phi^1;\\
\frac{1}{2}\varrho\frac{\partial U^2}{\partial t} + \varrho U^2\frac{\partial U^2}{\partial q^2} + g^{22}\frac{\partial p}{\partial q^2} &= \frac{1}{2}\Phi^2;\\
\frac{1}{2}\frac{\partial \varrho}{\partial t} + U^2\frac{\partial \varrho}{\partial q^2} + \varrho\frac{\partial U^2}{\partial q^2} &= \frac{1}{2}\Psi;\\
\frac{1}{2}\frac{\partial S}{\partial t} + U^2\frac{\partial S}{\partial q^2} &= 0.
\end{aligned}\right\} \tag{8.4.10}$$

Systems (8.4.9) and (8.4.10) are approximated by some difference scheme within the intervals $t \in [n\tau, (n + 1/2)\tau]$ and $[(n + 1/2)\tau, (n + 1)\tau]$, respectively.

Each of the systems (8.4.9) and (8.4.10) has the structure of a one-dimensional system. Namely, Eqs. (8.4.9a, b, c) are analogous to the one-dimensional equations of gas dynamics and can be solved with an implicit scheme which is based on vector iteration. Eq. (8.4.9b) is then solved by the method of running computation for U^2. A similar procedure can also be applied to system (8.4.10). This scheme of calculation is convenient in the case when a contact boundary in a movable system of coordinates represents the coordinate line (S. K. Godunov used movable networks of this type). In this case we can iterate along the channel crossing the boundary, because the transverse contravariant components of the velocity vector are continuous across the boundary (Fig. 13). This situation follows from the fact that a discontinuity of

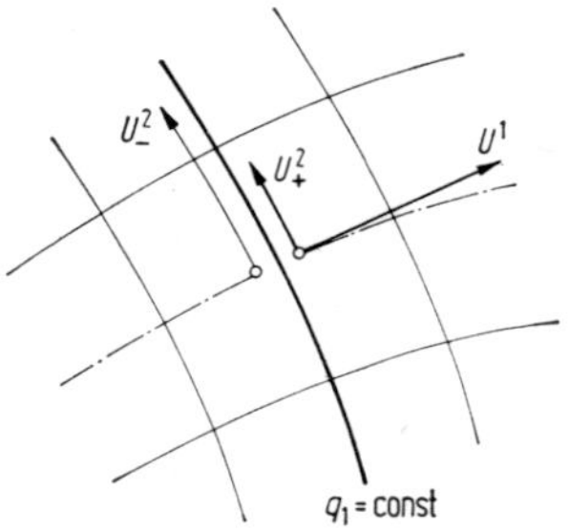

Fig. 13. U^1 is continuous across the contact surface; U^2 has a discontinuity

the velocity vector $U^+ - U^-$ at the contact boundary is parallel to the boundary itself.

8.5 The predictor-corrector method (method of correctors)

The predictor-corrector method can be considered as a modification of the method of fractional steps, although the predictor-corrector method has been used before in many problems.

As was demonstrated before, the predictor-corrector method can be used in the construction of the approximation correction scheme (Ch. 2). In the case of nonlinear equations it can be used for the reconstruction of conservative schemes. Although the predictor-corrector method has been used for ordinary differential equations for some time, its application to nonlinear equations with partial derivatives was suggested only recently by J. Douglas [75], S. K. Godunov, K. A. Semendyaev [76], S. K. Godunov [34], I. K. Yaushev and the author [35]. The application of the predictor-corrector method to equations of gas dynamics can be presented as follows. Let

$$\oint u\,dx - v(u)\,dt = \iint f\,dx\,dt \tag{8.5.1}$$

be a system of conservation laws presented in the vector form

$$u = \{u_1, \ldots, u_m\}, \qquad v = \{v_1, \ldots, v_m\}, \qquad f = \{f_1, \ldots, f_m\}.$$

System (8.5.1) corresponds to the conservative hyperbolic quasilinear system

$$\frac{\partial u}{\partial t} + \frac{\partial v(u)}{\partial x} = f(u). \tag{8.5.2}$$

Using a rectangular network, relations (8.5.2) are approximated by the difference scheme (Fig. 14):

$$\frac{u_i^{n+1} - u_i^n}{\tau} + \frac{v_{i+1}^* - v_{i-1}^*}{2h} = f(u_i^{**}). \tag{8.5.3}$$

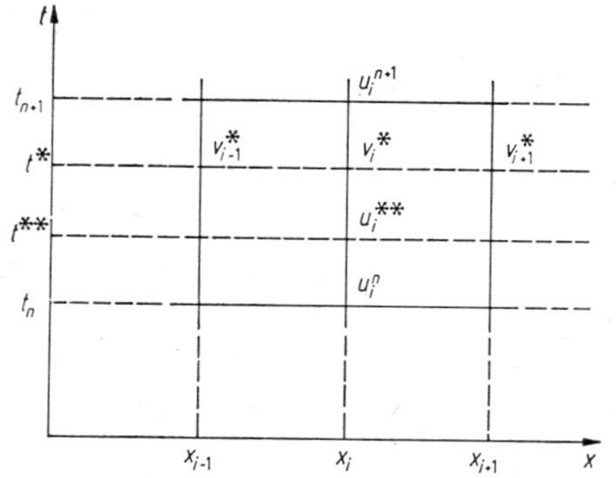

Fig. 14. Distribution of values u_i^n, v_i^*, u_i^{**}

It was assumed here that

$$u_i^n = u(x_i, t_n), \qquad v_i^* = v(x_i, t^*), \qquad u_i^{**} = u(x_i, t^{**}). \qquad (8.5.4)$$

Quantities v_i^*, u_i^{**} can be calculated with any scheme. Consider first a uniform scheme obtained by assuming that

$$v_i^* = \alpha v_i^{n+1} + \beta v_i^n = \alpha v(u_i^{n+1}) + \beta v(u_i^n),$$
$$u_i^{**} = u_i^* = \alpha u_i^{n+1} + \beta u_i^n, \quad \alpha \geq 0, \quad \beta \geq 0, \quad \alpha + \beta = 1. \qquad (8.5.5)$$

Scheme (8.5.3) takes the form

$$\frac{u_i^{n+1} - u_i^n}{\tau} + \frac{\alpha[v(u_{i+1}^{n+1}) - v(u_{i-1}^{n+1})]}{2h} + \frac{\beta[v(u_{i+1}^n) - v(u_{i-1}^n)]}{2h}$$
$$= f(u_i^*) = f(\alpha u_i^{n+1} + \beta u_i^n). \qquad (8.5.6)$$

For the realization of scheme (8.5.6) formulas (8.5.6) should first be linearized with the assumption that

$$v(u^{n+1}) = v(u^n) + \left\| \frac{\partial v^n}{\partial u} \right\| (u^{n+1} - u^n);$$
$$f(u^*) = f(u^n) + \left\| \frac{\partial f^n}{\partial u} \right\| (u^* - u^n). \qquad (8.5.7)$$

After this formulas (8.5.6) are solved by vector sweeps. It should be noted that after the introduction of Eq. (8.5.7) into Eq. (8.5.6) the scheme becomes non-conservative. In order to retain this property scheme (8.5.6) should be solved exactly and relations (8.5.7) considered as iterative

$$v(\overset{k+1}{u}) = v(\overset{k}{u}) + \left\| \frac{\partial \overset{k}{v}}{\partial u} \right\| (\overset{k+1}{u} - \overset{k}{u});$$
$$f(\overset{k+1}{u}) = f(\overset{k}{u}) + \left\| \frac{\partial \overset{k}{f}}{\partial u} \right\| (\overset{k+1}{u} - \overset{k}{u}), \qquad (8.5.8)$$

where k is the iteration index and $\overset{0}{u} = u^n$. Thus in the case of a uniform scheme it is necessary to combine the vector iterations with nonlinear iterations. The predictor-corrector method makes it possible to do this without iterations. An arbitrary implicit absolutely stable scheme can be used for the calculation of v^*, which does not have to be conservative and which does not require nonlinear iteration.

System (8.5.2) can be written as

$$\frac{\partial u}{\partial t} + A \frac{\partial u}{\partial x} = f; \qquad A = \left\| \frac{\partial v}{\partial u} \right\|. \qquad (8.5.9)$$

Then the following scheme can be used for determination of u^* and v^*:

$$\frac{u^* - u^n}{\tau} + A(u^n) \frac{\Delta_1 + \Delta_{-1}}{2h} u^* = f(u^n) + B(u^n)(u^* - u^n),$$
$$B = \left\| \frac{\partial f}{\partial u} \right\|. \qquad (8.5.10)$$

This scheme can be realized by means of vector sweeps and does not require iterations (see [34]).

After u^* is determined from Eq. (8.5.10) we determine $v^* = v(u^*)$. With scheme (8.5.3) we obtain a conservation scheme. It is absolutely stable for $t^* \geq (t_n + t_{n+1})/2$ and is of second order accuracy for $t^* = (t_n + t_{n+1})/2$. Another implicit scheme, for example, the majorant characteristic scheme, can be used instead of (8.4.10) (see [35]).

8.6 The equations of meteorology

G. I. Marchuk [77, 78, 97] developed the following implicit splitting scheme for the integration of the equations of meteorology:

$$\frac{\partial u^1}{\partial t} + u^1 \frac{\partial u^1}{\partial x^1} + u^2 \frac{\partial u^1}{\partial x^2} - l\, u^2 = -\frac{\partial H}{\partial x^1}, \tag{8.6.1 a}$$

$$\frac{\partial u^2}{\partial t} + u^1 \frac{\partial u^2}{\partial x^1} + u^2 \frac{\partial u^2}{\partial x^2} + l\, u^1 = -\frac{\partial H}{\partial x^2}, \tag{8.6.1 b}$$

$$\frac{\partial T}{\partial t} + u^1 \frac{\partial T}{\partial x^1} + u^2 \frac{\partial T}{\partial x^2} - k_1 T \frac{u^3}{p} = q, \tag{8.6.1 c}$$

$$\frac{\partial u^1}{\partial x^1} + \frac{\partial u^2}{\partial x^2} + \frac{\partial u^3}{\partial p} = 0, \tag{8.6.1 d}$$

$$T = -k_2\, p \frac{\partial H}{\partial p}, \tag{8.6.1 e}$$

where u^1, u^2, u^3 are velocity components along axes x^1, x^2, p; T is the temperature; H is the height of the isobaric surface $p = \text{const}$; q is the heat flow per unit mass; l is the Coriolis parameter; k_1 and k_2 are constants. System (8.6.1) is irregular, i.e., does not belong to the Cauchy-Kowalewski systems.

After the introduction of the vector function

$$u = \{u^1, u^2, T\}; \qquad f = \left\{-\frac{\partial H}{\partial x^1}, -\frac{\partial H}{\partial x^2}, k_1 \frac{T u^3}{p} + q\right\} \tag{8.6.2}$$

and the matrix

$$A = \begin{Vmatrix} 0 & -l & 0 \\ l & 0 & 0 \\ 0 & 0 & 0 \end{Vmatrix} \tag{8.6.3}$$

Eqs. (8.6.1 a, b, c) can be written

$$\frac{\partial u}{\partial t} + u^1 \frac{\partial u}{\partial x^1} + u^2 \frac{\partial u}{\partial x^2} + A\, u = f. \tag{8.6.4}$$

Subsystem (8.6.4) is a closed regular system of hyperbolic type for a given function f. It is solved by the method of splitting

$$\frac{u^{n+1/3} - u^n}{\tau} + \Lambda_1 u^{n+1/3} = 0, \tag{8.6.5 a}$$

$$\frac{u^{n+2/3} - u^{n+1/3}}{\tau} + \Lambda_2 u^{n+2/3} = 0, \tag{8.6.5 b}$$

$$\frac{u^{n+1} - u^{n+2/3}}{\tau} + A\, u^{n+1} = f^{n+1}, \tag{8.6.5 c}$$

where operators Λ_1, Λ_2 approximate the operators $u^1(\partial/\partial x^1)$, $u^2(\partial/\partial x^2)$, respectively. In order to close system (8.6.5), we need to determine f^{n+1}. For this purpose a difference approximation of Eqs. (8.6.1 d) and (8.6.1 e) is added to Eq. (8.6.5 c):

$$\left.\begin{aligned} &\sum_{k=1}^{3} \frac{\Delta_k + \Delta_{-k}}{2h_k} (u^k)^{n+1} = 0, \\ &T^{n+1} = -k_2 p \frac{\Delta_3 + \Delta_{-3}}{2h_3} H^{n+1}, \\ &h_1 = \Delta x_1, \quad h_2 = \Delta x_2, \quad h_3 = \Delta p. \end{aligned}\right\} \tag{8.6.6}$$

Eq. (8.6.5 c) together with (8.6.6) represent a closed system with respect to u^{n+1} and H^{n+1} (considering boundary conditions). When u^{n+1} is eliminated from Eqs. (8.6.5 c) and (8.6.6) it is possible to obtain an equation for H^{n+1} which is solved by iteration.

Chapter 9

General Definitions

In the preceding chapters several problems of mathematical physics were studied together with their solution, using the method of fractional steps. In this chapter an attempt is made to present a more general point of view and to discuss different methods of construction of schemes with fractional steps, as well as to define the method of fractional steps for a wide class of schemes.

9.1 General formulation of the method of splitting Validity of the method as determined by the elimination principle in the commutative case

In [79] the method of splitting (fractional steps) was formulated as a method for the construction of economical implicit schemes for a system of partial differential equations. In [79] only the two-layer systems in fractional steps were studied. In the paper of G. I. Marchuk and the author [51] presented at the All-Union Conference on Computational Mathematics (Moscow, 1965) and at IFIP (New York, 1965) this method was formulated for systems of differential equations and many-layer schemes in fractional steps. Consider a linear system of integro-differential equations with respect to an unknown vector function

$$\frac{\partial u}{\partial t} = \Omega u + f, \tag{9.1.1}$$

for which the Cauchy problem is correctly posed in some Banach space

$$u(x, 0) = u_0(x). \tag{9.1.2}$$

Let

$$\Omega = \Omega_1 + \Omega_2 + \cdots + \Omega_p \tag{9.1.3}$$

be the representation of integro-differential operator Ω as the sum of p operators $\Omega_1, \Omega_2, \ldots, \Omega_p$; the operators $\Omega_1, \ldots, \Omega_p$ are approximated by operators Λ_{ij} in such a way that the following approximate representations are valid[1]

$$\left.\begin{array}{ll} \Lambda_{10} + \Lambda_{11} & \sim \Omega_1; \\ \Lambda_{20} + \Lambda_{21} + \Lambda_{22} & \sim \Omega_2; \\ \dots\dots\dots\dots & \\ \Lambda_{p0} + \Lambda_{p1} + \cdots + \Lambda_{pp} & \sim \Omega_p. \end{array}\right\} \qquad (9.1.4)$$

The method of splitting is

$$\left.\begin{array}{l} \dfrac{u^{n+1/p} - u^n}{\tau} = \Lambda_{10} u^n + \Lambda_{11} u^{n+1/p} + F_1; \\ \dfrac{u^{n+2/p} - u^{n+1/p}}{\tau} = \Lambda_{20} u^n + \Lambda_{21} u^{n+1/p} + \Lambda_{22} u^{n+2/p} + F_2; \\ \dots\dots\dots\dots\dots\dots \\ \dfrac{u^{n+1} - u^{n+(p-1)/p}}{\tau} = \Lambda_{p0} u^n + \Lambda_{p1} u^{n+1/p} + \cdots + \Lambda_{pp} u^{n+1} + F_p, \end{array}\right\} \qquad (9.1.5)$$

where

$$F_s = \Lambda_s f; \quad \sum_{s=1}^{p} \Lambda_s \sim E. \qquad (9.1.6)$$

If

$$\Lambda_{sr} = 0, \quad r < s - 1, \qquad (9.1.7)$$

then each of the schemes (9.1.5) is a two-layer scheme. It is not difficult to define the convergence conditions of the difference scheme (9.1.5) to Eq. (9.1.1), provided all difference operators Λ_{ij}, Λ_i are commutative.

Consider for simplicity the case of uniform two-layer schemes ($f = 0$, $F_s = 0$, $\Lambda_{sr} = 0$, $r < s - 1$)

$$\frac{u^{n+k/p} - u^{n+(k-1)/p}}{\tau} = \Lambda_{kk-1} u^{n+(k-1)/p} + \Lambda_{kk} u^{n+k/p}, \quad k = 1, \dots, p. \qquad (9.1.8)$$

Construct a corresponding scheme in whole steps. Eqs. (9.1.8) can be rewritten in the form

$$A_k u^{n+k/p} = B_k u^{n+(k-1)/p}, \qquad (9.1.9)$$

where

$$A_k = E - \tau \Lambda_{kk}; \quad B_k = E + \tau \Lambda_{kk-1}, \quad k = 1, \dots, p.$$

Using the method of elimination suggested by J. Douglas [12] and also the commutative property of operators A_k and B_k the following scheme

[1] Operators Λ_{ij} can be of arbitrary structure both difference and integro-differential.

in whole steps is obtained

$$A\, u^{n+1} = B\, u^n; \qquad A = A_1 \dots A_p, \qquad B = B_1 \dots B_p. \tag{9.1.10}$$

Expansion of operators A and B in series of τ results in

$$\left.\begin{aligned} A &= E - \tau(\Lambda_{11} + \Lambda_{22} + \cdots + \Lambda_{pp}) + \\ &\quad + \tau^2(\Lambda_{11}\Lambda_{22} + \cdots + \Lambda_{p-1\,p-1}\Lambda_{pp}) + \cdots + \\ &\quad + (-1)^p \tau^p \Lambda_{11}\Lambda_{22} \dots \Lambda_{p-1\,p-1}\Lambda_{pp}; \\ B &= E + \tau(\Lambda_{10} + \Lambda_{21} + \cdots + \Lambda_{pp-1}) + \\ &\quad + \tau^2(\Lambda_{10}\Lambda_{21} + \cdots + \Lambda_{p-1\,p-2}\Lambda_{pp-1}) + \cdots + \\ &\quad + \tau^p \Lambda_{10}\Lambda_{21} \dots \Lambda_{pp-1}. \end{aligned}\right\} \tag{9.1.11}$$

Using Eq. (9.1.11), the scheme (9.1.10) can be transformed into

$$\frac{u^{n+1} - u^n}{\tau} = \sum_{k=1}^{p} (\Lambda_{kk-1}\, u^n + \Lambda_{kk}\, u^{n+1}) + \tau\, \Phi, \tag{9.1.12}$$

where

$$\begin{aligned} \Phi &= \sum_{i<j} (\Lambda_{i\,i-1}\Lambda_{j\,j-1}\, u^n - \Lambda_{i\,i}\Lambda_{j\,j}\, u^{n+1}) + \tau \sum_{i<j<k} (\Lambda_{i\,i-1}\Lambda_{j\,j-1}\Lambda_{kk-1}\, u^n + \\ &\quad + \Lambda_{i\,i}\Lambda_{j\,j}\Lambda_{kk}\, u^{n+1}) + \cdots + \tau^{p-2}[\Lambda_{10}\Lambda_{21} \dots \Lambda_{pp-1}\, u^n + \\ &\quad + (-1)^{p-1}\Lambda_{11}\Lambda_{22} \dots \Lambda_{pp}\, u^{n+1}]. \end{aligned} \tag{9.1.13}$$

Eq. (9.1.12) proves the consistency of the two-layer scheme (9.1.9) because of Eqs. (9.1.4) and (9.1.7)

$$\sum_{k=1}^{p} (\Lambda_{kk-1} + \Lambda_{kk}) \sim \sum_{k=1}^{p} \Omega_k = \Omega. \tag{9.1.14}$$

Now the conditions of correctness of schemes (9.1.9) and (9.1.10) have to be found. Eqs. (9.1.9) can be expressed as follows

$$u^{n+k/p} = C_k\, u^{n+(k-1)/p}, \qquad C_k = A_k^{-1} B_k, \qquad k = 1, \dots, p. \tag{9.1.15}$$

The scheme in whole steps is

$$u^{n+1} = C\, u^n, \qquad C = C_p\, C_{p-1} \dots C_1 = A^{-1} B. \tag{9.1.16}$$

Scheme (9.1.9) is uniformly correct if

$$\|C\| = \|C_p\, C_{p-1} \dots C_1\| \le 1 + \text{const}\, \tau. \tag{9.1.17}$$

It is obvious that the stability of each scheme (9.1.9) in fractional steps is not a necessary condition.

Using a known theorem (see Ch. 1), convergence occurs if

$$\|A^{-1}\| \le M, \qquad \|C\| \le 1 + N\tau, \tag{9.1.18}$$

where constants M and N do not depend on n and τ.

9.2 Validity of the method of splitting in the non-commutative case

In the general case of the non-commutative operators the convergence criteria are analogous but the proof is more complex. Changes of the method of elimination in the case of non-commutative operators are shown below for the two-layer scheme (9.1.9) with two fractional steps ($p = 2$).

Since the operators A_k and B_k are non-commutative, the usual method of elimination is not applicable. Multiply the first of Eqs. (9.1.9) by $\bar{B}_2$ and the second by $\bar{A}_1$ (where operators $\bar{A}_1$ and $\bar{B}_2$ are still undetermined). Addition results in

$$\bar{A}_1 A_2 u^{n+1} - \bar{B}_2 B_1 u^n = (\bar{A}_1 B_2 - \bar{B}_2 A_1) u^{n+1/2}. \qquad (9.2.1)$$

Assume that

$$\bar{A}_1 = A_1 + \tau^s a_1; \qquad \bar{B}_2 = B_2 + \tau^s b_2 \qquad (9.2.2)$$

with undetermined operators a_1, b_2 and exponent s. It follows from this expression that

$$\bar{A}_1 B_2 - \bar{B}_2 A_1 = \tau^2 (\Lambda_{21} \Lambda_{11} - \Lambda_{11} \Lambda_{21}) + \tau^s (a_1 B_2 - b_2 A_1). \qquad (9.2.3)$$

Assuming that

$$\bar{A}_1 B_2 - \bar{B}_2 A_1 = 0, \qquad (9.2.4)$$

we obtain

$$s = 2, \qquad (9.2.5)$$

$$b_2 A_1 - a_1 B_2 = \Lambda_{21} \Lambda_{11} - \Lambda_{11} \Lambda_{21}. \qquad (9.2.6)$$

If operators a_1 and b_2 satisfy condition (9.2.6), then Eq. (9.2.1) is

$$\frac{u^{n+1} - u^n}{\tau} = (\Lambda_{11} + \Lambda_{22}) u^{n+1} + (\Lambda_{10} + \Lambda_{21}) u^n + \tau \Phi, \qquad (9.2.7)$$

$$\Phi = -\Lambda_{11} \Lambda_{22} u^{n+1} + \Lambda_{21} \Lambda_{10} u^n + b_2 B_1 u^n - a_1 A_2 u^{n+1}. \qquad (9.2.8)$$

Assume also for simplicity that $a_1 = 0$. Then from Eq. (9.2.6) we have

$$b_2 = (\Lambda_{21} \Lambda_{11} - \Lambda_{11} \Lambda_{21}) A_1^{-1}. \qquad (9.2.9)$$

From this Φ takes the form

$$\Phi = -\Lambda_{11}\Lambda_{22}u^{n+1} + \Lambda_{21}\Lambda_{10}u^n + (\Lambda_{21}\Lambda_{11} - \Lambda_{11}\Lambda_{21}) A_1^{-1} B_1 u^n. \qquad (9.2.10)$$

Using condition (9.1.14) we see that the scheme (9.1.7) is consistent with Eq. (9.1.1) if the value

$$(\Lambda_{21} \Lambda_{11} - \Lambda_{11} \Lambda_{21}) A_1^{-1} B_1 f \qquad (9.2.11)$$

is finite for any sufficiently smooth function f.

It is shown below that with certain restrictions on the A operator, the function $A^{-1}F$ possesses the same smoothness as F, i.e., the boundedness of $A^{-1}F$ and of its finite differences up to order q for any small h, follows from the boundedness of F and its finite differences up to order q.

Let A be the difference operator (not necessarily finite) of the general type

$$A : A f = \sum_{\alpha} a_{\alpha_1 \ldots \alpha_m} T_1^{\alpha_1} \ldots T_m^{\alpha_m} f, \tag{9.2.12}$$

where the sums with respect to $\alpha_1, \ldots, \alpha_m$ are carried out from

$$\alpha_s = -N_s \quad \text{to} \quad \alpha_s = N_s, \qquad s = 1, \ldots, m.$$

Let $\Delta_s = T_s - E$ be the operator of the first right-hand difference. The following operators should also be considered: $\delta_i A$, $\delta_{ij} A, \ldots,$ $\delta_{i_1 \ldots i_q} A$

$$\left.\begin{aligned} \delta_i A &= \sum_{\alpha} (\Delta_i a_{\alpha_1 \ldots \alpha_m}) T_1^{\alpha_1} \ldots T_m^{\alpha_m}; \\ \delta_{i_1 i_2} A &= \sum_{\alpha} (\Delta_{i_1} \Delta_{i_2} a_{\alpha_1 \ldots \alpha_m}) T_1^{\alpha_1} \ldots T_m^{\alpha_m}; \\ &\ldots\ldots\ldots\ldots\ldots\ldots\ldots\ldots \\ \delta_{i_1 \ldots i_q} A &= \sum_{\alpha} (\Delta_{i_1} \ldots \Delta_{i_q} a_{\alpha_1 \ldots \alpha_m}) T_1^{\alpha_1} \ldots T_m^{\alpha_m}. \end{aligned}\right\} \tag{9.2.13}$$

The following relations can easily be verified

$$\left.\begin{aligned} \Delta_i (A f) &= (A + \delta_i A) \Delta_i f + (\delta_i A) f; \\ \Delta_i \Delta_j (A f) &= (A + \delta_i A + \delta_j A + \delta_{ij} A) \Delta_i \Delta_j f + \\ &\quad + (\delta_i A + \delta_{ij} A) \Delta_j f + (\delta_j A + \delta_{ij} A) \Delta_i f + (\delta_{ij} A) f. \end{aligned}\right\} \tag{9.2.14}$$

It is clear that analogous equalities can be obtained for the left-hand difference and also for differences of any order.

Lemma 1. *If we assume*

$$\delta_i A = h_i A_i; \; \delta_{i_1 i_2} A = h_{i_1} h_{i_2} A_{i_1 i_2}, \ldots, \delta_{i_1 \ldots i_q} A = h_{i_1} \ldots h_{i_q} A_{i_1 \ldots i_q}, \tag{9.2.15}$$

and operators $A, A_i, A_{i_1 i_2}, \ldots, A_{i_1 \ldots i_q}$ *satisfy conditions*

$$\begin{aligned} &\|A\| \le M_1; \quad \|A^{-1}\| \le M_1; \quad \|A_i\| \le M_1; \\ &\|A_{i_1 i_2}\| \le M_1, \ldots; \quad \|A_{i_1 \ldots i_q}\| \le M_1, \end{aligned} \tag{9.2.16}$$

where M_1 *does not depend on* $h_1, \ldots, h_m$, *then*

$$\begin{aligned} &\delta_i A^{-1} = h_i B_i, \quad \delta_{i_1 i_2} A^{-1} = h_{i_1} h_{i_2} B_{i_1 i_2}; \\ &\ldots\ldots\ldots\ldots\ldots\ldots\ldots\ldots \\ &\delta_{i_1 \ldots i_q} A^{-1} = h_{i_1} \ldots h_{i_q} B_{i_1 \ldots i_q}, \end{aligned} \tag{9.2.17}$$

where operators $B_i, B_{i_1 i_2}, \ldots, B_{i_1 \ldots i_q}$ *satisfy the relations*

$$\|B_i\| \le M_2; \quad \|B_{i_1 i_2}\| \le M_2, \ldots; \quad \|B_{i_1 \ldots i_q}\| \le M_2 \tag{9.2.18}$$

and M_2 *does not depend on* $h_1, \ldots, h_m$.

Proof. If the operator δ_i is applied to the equality $A\,A^{-1} = E$ (E is the unit matrix), then we have

$$(A + \delta_i A)\, \delta_i A^{-1} + \delta_i A \cdot A^{-1} = 0. \tag{9.2.19}$$

It follows that

$$\begin{aligned} \delta_i A^{-1} &= -(A + h_i A_i)^{-1}\, \delta_i A \cdot A^{-1} \\ &= -h_i (A + h_i A_i)^{-1} A_i A^{-1} = -h_i B_i, \end{aligned} \tag{9.2.20}$$

and B_i satisfies Eq. (9.2.18). □

The inequalities (9.2.18) for $B_{i_1 i_2}, \ldots, B_{i_1 i_2 \ldots i_q}$ are proved in an analogous way.

Lemma 2. *If* (a) *the operator A satisfies conditions* (9.2.15) *and* (9.2.16) *of Lemma* 1;

(b) $\|f\| \le M_3$; $\|\Delta_i f\| \le h_i M_3, \ldots$; $\|\Delta_{i_1}\Delta_{i_2} \ldots \Delta_{i_q} f\| \le h_{i_1} \ldots h_{i_q} M_3$,

where the constant M_3 does not depend on h_i, then the following bounds can be established

$$\begin{aligned} \|A^{-1} f\| \le M_4; \quad \|\Delta_i (A^{-1} f)\| &\le h_i M_4, \ldots; \\ \|\Delta_{i_1} \ldots \Delta_{i_q} (A^{-1} f)\| &\le h_{i_1} \ldots h_{i_q} M_4; \end{aligned} \tag{9.2.21}$$

where M_4 does not depend on h_i.

Proof. The inequality $\|A^{-1} f\| \le M_4$ is obvious. Let us estimate the quantity $\Delta_i (A^{-1} f)$. Using Eq. (9.2.14) and applying it to operator A^{-1} we have

$$\Delta_i (A^{-1} f) = (A^{-1} + \delta_i A^{-1})\, \Delta_i f + \delta_i A^{-1} f \tag{9.2.22}$$

or in the notation (9.2.17)

$$\frac{\Delta_i}{h_i} (A^{-1} f) = (A^{-1} + h_i B_i) \frac{\Delta_i}{h_i} f + B_i f. \tag{9.2.23}$$

Using (9.2.16) and (9.2.18) of Lemma 1 we find that

$$\begin{aligned} \left\| \frac{\Delta_i}{h_i} (A^{-1} f) \right\| &\le [\|A^{-1}\| + h_i \|B_i\|] \left\| \frac{\Delta_i}{h_i} f \right\| + \|B_i\|\, \|f\| \\ &\le (M_1 + h_i M_2) \left\| \frac{\Delta_i}{h_i} f \right\| + M_2 \|f\|. \end{aligned} \tag{9.2.24}$$

Thus Lemma 2 for $\Delta_i (A^{-1} f)$ is proved. □

The result for $\Delta_{i_1} \ldots \Delta_{i_q} (A^{-1} f)$ is proved in an analogous way.

When Lemmas 1 and 2 are applied to Eq. (9.2.11) we see that, for sufficiently smooth behavior of the coefficients of Eq. (9.2.1) and of its solution $u(x, t)$, the required consistency can be achieved.

The idea of this proof of consistency was suggested by the author in his paper presented at the IVth All-Union Mathematical Conference [28], and in more detail in his other paper [80], and also in the work of

Yu. E. Boyarintsev [31]. Many works have been devoted to application of the method of fractional steps (with the use of *a priori* estimates) for a wide class of equations with variable coefficients. The following works should be mentioned in this connection: M. Lees [82, 83], E. G. D'yakonov [21 – 24, 95], A. A. Samarskii [84, 85, 41, 63].

9.3 The method of approximate factorization of an operator

The method of exact factorization of the operator at the upper layer, which, in the case of the equation of heat conduction, leads to a scheme analogous to the splitting scheme, is formulated in the work of G. A. Baker and T. A. Oliphant [16]. However, the method of exact factorization turns out to be unacceptable for equations with variable coefficients. In the author's work [18] the idea of an approximate factorization of an operator was defined, which is applicable to equations with variable coefficients. The method of approximate factorization (the method of a split operator in the terminology of E. G. D'yakonov) was studied and developed for a wide class of parabolic and hyperbolic equations in the works of E. G. D'yakonov [21 – 24, 95].

A general formulation of the method from the point of view of G. I. Marchuk and the author [51] is presented below.

Let

$$\frac{u^{n+1} - u^n}{\tau} = \Lambda_1 u^{n+1} + \Lambda_0 u^n + F_n \tag{9.3.1}$$

be the uniform difference approximation of Eq. (9.1.1). The function F_n approximates f and in the case of multi-layer schemes also contains the result of application of difference operators to u^{n-1}, u^{n-2}, etc. We confine attention to the two-layer approximation in the definition of the method of approximate factorization.

Let

$$\begin{aligned} \Lambda_1 &= \Lambda_{11} + \Lambda_{12} + \cdots + \Lambda_{1p}; \\ \Lambda_0 &= \Lambda_{01} + \Lambda_{02} + \cdots + \Lambda_{0q} \end{aligned} \tag{9.3.2}$$

be a representation of the upper and lower operators Λ_1, Λ_0 as a sum of operators of simpler structure. The following approximation relations are valid

$$\begin{aligned} (E - \tau \Lambda_{11})(E - \tau \Lambda_{12}) \ldots (E - \tau \Lambda_{1p}) &\sim E - \tau \Lambda_1; \\ (E + \tau \Lambda_{01})(E + \tau \Lambda_{02}) \ldots (E + \tau \Lambda_{0q}) &\sim E + \tau \Lambda_0. \end{aligned} \tag{9.3.3}$$

Using Eq. (9.3.3) the scheme (9.3.1) can be replaced by the factorized scheme

$$\begin{aligned} &(E - \tau \Lambda_{11})(E - \tau \Lambda_{12}) \ldots (E - \tau \Lambda_{1p}) u^{n+1} \\ &\quad = (E + \tau \Lambda_{01})(E + \tau \Lambda_{02}) \ldots (E + \tau \Lambda_{0q}) u^n + \tau F_n. \end{aligned} \tag{9.3.4}$$

It is not too difficult to establish that scheme (9.3.4) is consistent with Eq. (9.1.1). In fact, expansion of Eq. (9.3.4) with respect to τ results in

$$\frac{u^{n+1}-u^n}{\tau} = (\Lambda_{11}+\Lambda_{12}+\cdots+\Lambda_{1p})\,u^{n+1}+ \\ +(\Lambda_{01}+\Lambda_{02}+\cdots+\Lambda_{0q})\,u^n+F_n+\tau\,\Phi, \qquad (9.3.5)$$

where

$$\Phi=\Big[\sum_{i<j}\Lambda_{0i}\Lambda_{0j}+\tau\sum_{i<j<k}\Lambda_{0i}\Lambda_{0j}\Lambda_{0k}+\cdots+\tau^{q-2}\Lambda_{01}\Lambda_{02}\ldots\Lambda_{0q}\Big]u_n+ \\ +\Big[-\sum_{i<j}\Lambda_{1i}\Lambda_{1j}+\tau\sum_{i<j<k}\Lambda_{1i}\Lambda_{1j}\Lambda_{1k}+\cdots+ \\ +(-1)^{p-1}\tau^{p-2}\Lambda_{11}\ldots\Lambda_{1p}\Big]u^{n+1}. \qquad (9.3.6)$$

Consistency follows from Eqs. (9.3.5) and (9.3.6).

Thus, as opposed to the method of splitting, consistency is achieved in the case of approximate factorization. Notice that the uniform scheme (9.3.1) and the representation (9.3.2) do not define exactly the same scheme. For example, scheme

$$(E-\tau\Lambda_{1p})(E-\tau\Lambda_{1p-1})\ldots(E-\tau\Lambda_{11})\,u^{n+1} \\ =(E+\tau\Lambda_{0q})(E+\tau\Lambda_{0q-1})\ldots(E+\tau\Lambda_{01})\,u^n+\tau F_n \qquad (9.3.4\,\text{a})$$

is not equivalent to scheme (9.3.4) in the general case of non-commutative operators Λ_{1s} and Λ_{0s} and becomes equivalent to it only in the case of commutative operators.

Let us now analyze stability, which is generally more difficult than for the method of splitting. Assume for simplicity that $f=0$. Eq. (9.3.4) is written in the form

$$u^{n+1}=(E-\tau\Lambda_{1p})^{-1}\ldots(E-\tau\Lambda_{11})^{-1}(E+\tau\Lambda_{01})\ldots(E+\tau\Lambda_{0q})\,u^n. \qquad (9.3.7)$$

If for the step operator

$$C=(E-\tau\Lambda_{1p})^{-1}\ldots(E-\tau\Lambda_{11})^{-1}(E+\tau\Lambda_{01})\ldots(E+\tau\Lambda_{0q}) \qquad (9.3.8)$$

we satisfy the condition

$$\|C\|\le 1+\text{const}\,\tau, \qquad (9.3.9)$$

then scheme (9.3.7) is uniformly correct.

The estimate of operator C from Eq. (9.3.8) is much more difficult. In the case when $p=q$ and the operators are commutative we have

$$\left.\begin{aligned} C&=(E-\tau\Lambda_{1p})^{-1}(E+\tau\Lambda_{0p})(E-\tau\Lambda_{1p-1})^{-1}(E+\tau\Lambda_{0p-1})\ldots \\ &\quad\ldots(E-\tau\Lambda_{11})^{-1}(E+\tau\Lambda_{01})=C_pC_{p-1}\ldots C_1, \\ C_s&=(E-\tau\Lambda_{1s})^{-1}(E+\tau\Lambda_{0s}). \end{aligned}\right\} \qquad (9.3.10)$$

The scheme of approximate factorization (9.3.4) is equivalent to the scheme of splitting in this case:

$$\left.\begin{aligned}
&\frac{u^{n+1/p}-u^n}{\tau}=\Lambda_{11}u^{n+1/p}+\Lambda_{01}u^n,\\
&\frac{u^{n+2/p}-u^{n+1/p}}{\tau}=\Lambda_{12}u^{n+2/p}+\Lambda_{02}u^{n+1/p},\\
&\dots\dots\dots\dots\dots\dots\dots\dots\dots\\
&\frac{u^{n+1}-u^{n+(p-1)/p}}{\tau}=\Lambda_{1p}u^{n+1}+\Lambda_{0\,p-1}u^{n+(p-1)/p}+H_n;\\
&H_n=A_{p-1}^{-1}\dots A_1^{-1}F_n.
\end{aligned}\right\}\qquad(9.3.11)$$

In fact, by eliminating successively $u^{n+1/p}, u^{n+2/p}, \dots, u^{n+(p-1)/p}$ from Eq. (9.3.11) we obtain the scheme

$$u^{n+1}=C_p\,C_{p-1}\dots C_1\,u^n+\tau\,A_p^{-1}\dots A_1^{-1}F_n.\qquad(9.3.12)$$

In the general case when the operators are non-commutative and $p \neq q$, the scheme (9.3.4) is equivalent to the following scheme in fractional steps:

$$\left.\begin{aligned}
&(E-\tau\Lambda_{11})u^{n+1/p}=(E+\tau\Lambda_{01})(E+\tau\Lambda_{02})\dots\\
&\qquad\dots(E+\tau\Lambda_{0q})u^n+\tau F_n;\\
&(E-\tau\Lambda_{12})u^{n+2/p}=u^{n+1/p},\dots;\ (E-\tau\Lambda_{1p})u^{n+1}=u^{n+(p-1)/p}.
\end{aligned}\right\}\qquad(9.3.13)$$

The latter in turn is equivalent to

$$\left.\begin{aligned}
&\frac{u^{n+1/p}-u^n}{\tau}=\Lambda_{11}u^{n+1/p}+\Omega\,u^n+F_n,\\
&\frac{u^{n+2/p}-u^{n+1/p}}{\tau}=\Lambda_{12}u^{n+2/p},\\
&\dots\dots\dots\dots\dots\dots\dots\dots\dots\\
&\frac{u^{n+1}-u^{n+(p-1)/p}}{\tau}=\Lambda_{1p}u^{n+1};\\
&\Omega=\frac{1}{\tau}[(E+\tau\Lambda_{01})(E+\tau\Lambda_{02})\dots(E+\tau\Lambda_{0q})-E].
\end{aligned}\right\}\qquad(9.3.14)$$

Thus, the realization of Eq. (9.3.14) is related to a successive conversion of the operators $(E-\tau\Lambda_{11}), (E-\tau\Lambda_{12}), \dots, (E-\tau\Lambda_{1p})$ of much simpler structure, as compared with operator $(E-\tau\Lambda_1)$, and this strongly simplifies the algorithm. It is obvious that the realization of the factorized scheme with fractional steps (the scheme of splitting) is not unique. In general, the factorized scheme (9.3.4) can be considered as some canonical scheme which is equivalent to different two-layer schemes of fractional steps.

Consider now the scheme of approximate factorization in the case of multi-layer schemes

$$A_1u^{n+1}+A_0u^n+A_{-1}u^{n+1}+\dots+A_{-p+1}u^{n-p+1}+f_n=0.\qquad(9.3.15)$$

Such schemes could be the results of uniform approximation of both system (9.1.1) and of a system of more complex form

$$B_1 \frac{\partial u}{\partial t} + B_2 \frac{\partial^2 u}{\partial t^2} + \cdots + B_p \frac{\partial^p u}{\partial t^p} = \Omega u + f, \qquad (9.3.16)$$

where $B_1, B_2, \ldots, B_p$ are linear operators.

Let operators A_s be represented as follows

$$A_s = E + \tau^{\alpha_s} \Lambda_s, \quad s = 1, 0, -1, \ldots, -p+1, \qquad (9.3.17)$$

where Λ_s represents space operators and

$$\Lambda_s = \Lambda_{s1} + \cdots + \Lambda_{s q_s}, \quad s = 1, \ldots, -p+1, \qquad (9.3.18)$$

is a representation of operators Λ_s as the sum of operators of simpler structure (generally speaking). Then the factorized scheme which corresponds to Eqs. (9.3.15) and (9.3.18) is approximately

$$\bar{A}_1 u^{n+1} + \bar{A}_0 u^n + \bar{A}_{-1} u^{n-1} + \cdots + \bar{A}_{-p+1} u^{n-p+1} + f_n = 0, \qquad (9.3.19)$$

where

$$\bar{A}_s = (E + \tau^{\alpha_s} \Lambda_{s1}) \ldots (E + \tau^{\alpha_s} \Lambda_{s q_s}) = A_s + \tau^{2\alpha_s} \Phi_s; \qquad (9.3.20)$$

$$\Phi_s = \sum_{i<j} \Lambda_{si} \Lambda_{sj} + \cdots + (\tau^{\alpha_s})^{q_s - 2} \Lambda_{s1} \ldots \Lambda_{s q_s}. \qquad (9.3.21)$$

Consistency of Eq. (9.3.19) is proved, as usual, by its expansion into series with respect to τ. The main difficulty here is the study of stability of this scheme.

In conclusion it should be said here that the principal point in the construction of the above schemes is an approximate factorization of the upper operator because this permits the simplification of their realization. The factorization of lower operators has a corresponding character and it guarantees the stability and accuracy of the scheme. Therefore, operators at the lower layers can be sought in the more general form

$$A_s = A_s + \tau^{\varrho_s} \Phi_s = E + \tau^{\alpha_s} \Lambda_s + \tau^{\varrho_s} \Phi_s, \qquad (9.3.22)$$

where operators Φ_s are of arbitrary structure. Here the method of fractional steps has the character of the method of undetermined operators (or coefficients). The theory of schemes of general structure has been developed very little at the present time.

9.4 The method of stabilizing corrections

The method of stabilizing corrections which was introduced by J. Douglas and H. H. Rachford [12] and formulated in its general form by J. Douglas and J. E. Gunn [86] is a very general and effective method for the construction of schemes with fractional steps.

Let

$$\frac{u^{n+1} - u^n}{\tau} = \Lambda\, u^{n+1} + F_n, \tag{9.4.1}$$

where

$$F_n = A_0\, u^n + A_{-1}\, u^{n-1} + \cdots + A_{-q+1}\, u^{n-q+1} + f_n \tag{9.4.2}$$

is the initial uniform scheme. Assume that

$$\Lambda = \Lambda_1 + \Lambda_2 + \cdots + \Lambda_p. \tag{9.4.3}$$

The following scheme in fractional steps is proposed for schemes (9.4.1), (9.4.2) and the representation (9.4.3):

$$\left.\begin{aligned} \frac{u^{n+1/p} - u^n}{\tau} &= \Lambda_1 (u^{n+1/p} - u^n) + \Lambda\, u^n + F_n, \\ \frac{u^{n+2/p} - u^{n+1/p}}{\tau} &= \Lambda_2 (u^{n+2/p} - u^n), \\ &\cdots\cdots\cdots\cdots \\ \frac{u^{n+1} - u^{n+(p-1)/p}}{\tau} &= \Lambda_p (u^{n+1} - u^n). \end{aligned}\right\} \tag{9.4.4}$$

By eliminating successively $u^{n+1/p}, u^{n+2/p}, \ldots, u^{n+(p-1)/p}$ an equivalent factorized scheme in whole steps is obtained

$$\begin{aligned} &(E - \tau\Lambda_1)\ldots(E - \tau\Lambda_p)\, u^{n+1} = \tau\Lambda u^n + \tau F_n + (E - \tau\Lambda_1)\, u^n - \\ &\quad - \tau (E - \tau\Lambda_1)\Lambda_2 u^n - \cdots - \tau (E - \tau\Lambda_1)\ldots(E - \tau\Lambda_{p-1})\Lambda_p u^n. \end{aligned} \tag{9.4.5}$$

After algebraic transformation we have

$$\frac{u^{n+1} - u^n}{\tau} = \Lambda\, u^{n+1} + F_n - \tau^2\, \Phi \left(\frac{u^{n+1} - u^n}{\tau}\right), \tag{9.4.6}$$

$$\Phi = \sum_{i<j} \Lambda_i \Lambda_j - \tau \sum_{i<j<k} \Lambda_i \Lambda_j \Lambda_k + \cdots + (-1)^p \tau^{p-2} \Lambda_1 \Lambda_2 \ldots \Lambda_p. \tag{9.4.7}$$

It follows from Eqs. (9.4.6) and (9.4.7) that the scheme of stabilizing corrections preserves the approximation of the initial scheme (9.4.1) and (9.4.2) with an accuracy up to $O(\tau^2)$.

Let us now analyze the stability of the scheme of stabilizing corrections.

Assume, for simplicity, that u is a scalar function and operators Λ_s and A_s are commutative. Then using the stability analysis of Eq. (9.4.1) we find that

$$\frac{\varrho_1 - 1}{\tau} = \lambda\, \varrho_1 + g(\varrho_1), \tag{9.4.8}$$

where ϱ_1 is the amplification factor of scheme (9.4.1); λ is an eigenvalue of operator Λ; then

$$g(\varrho_1) = a_0 + a_{-1} \frac{1}{\varrho_1} + \cdots + a_{-q+1} \frac{1}{\varrho_1^{q-1}}$$

is the growth of error which corresponds to the operator

$$A_0 u^n + A_{-1} u^{n-1} + \cdots + A_{-q+1} u^{n-q+1},$$

$a_0, a_{-1}, \ldots, a_{-q+1}$ are eigenvalues of the operators

$$A_0, A_{-1}, \ldots, A_{-q+1}.$$

Analogously, for scheme (9.4.6) we find that

$$\frac{\varrho - 1}{\tau} = \lambda \varrho + g(\varrho) - \tau^2 \varphi\left(\frac{\varrho - 1}{\tau}\right), \tag{9.4.9}$$

where $\varphi[(\varrho - 1)/\tau]$ is the error growth which corresponds to the operator $\Phi[(u^{n+1} - u^n)/\tau]$. For $\varphi[(\varrho - 1)/\tau]$ we have

$$\varphi\left(\frac{\varrho - 1}{\tau}\right) = \frac{a}{\tau}(\varrho - 1), \tag{9.4.10}$$

where

$$a = \sum_{i<j} \lambda_i \lambda_j - \tau \sum_{i<j<k} \lambda_i \lambda_j \lambda_k + \cdots + (-1)^p \tau^{p-2} \lambda_1 \ldots \lambda_p; \tag{9.4.11}$$

$\lambda_1, \ldots, \lambda_p$ are eigenvalues of operators $\Lambda_1, \ldots, \Lambda_p$. Thus, the following expressions are obtained for the amplification factors ϱ_1 and ϱ:

$$\varrho_1 = \frac{1 + \tau g(\varrho_1)}{1 - \lambda \tau}; \qquad \varrho = \frac{1 + \tau g(\varrho) + a \tau^2}{1 - \lambda \tau + a \tau^2}. \tag{9.4.12}$$

In the case of two-layer schemes $g(x)$ does not depend on x and the following theorem holds.

If λ_i is negative (it follows from this that a is positive) then stability of Eq. (9.4.6) follows from the stability of Eq. (9.4.1). A more detailed analysis can be found in [86].

9.5 The method of approximation corrections

The method of approximation corrections is distinguished by the fact that the maximum stability is achieved in the preliminary steps and an accurate scheme is constructed at the last step. Let

$$\frac{u^{n+1} - u^n}{\tau} = \Lambda \frac{u^{n+1} + u^n}{2} + f_n \tag{9.5.1}$$

be a uniform two-layer scheme which approximates Eq. (9.1.1). Assume that

$$\Lambda = \Lambda_1 + \cdots + \Lambda_p. \tag{9.5.2}$$

The scheme of approximation corrections which corresponds to Eqs. (9.5.1) and (9.5.2) is

$$\begin{aligned} \frac{u^{n+1/2p} - u^n}{\tau/2} &= \Lambda_1 u^{n+1/2p}, \\ &\cdots\cdots\cdots \\ \frac{u^{n+p/2p} - u^{n+(p-1)/2p}}{\tau/2} &= \Lambda_p u^{n+p/2p} = \Lambda_p u^{n+1/2}, \end{aligned} \tag{9.5.3}$$

$$\frac{u^{n+1} - u^n}{\tau} = \Lambda u^{n+1/2} + F_n, \tag{9.5.4}$$

where F_n is some approximation of f_n. By eliminating successively $u^{n+1/2p}, \ldots, u^{n+(p-1)/2p}$ we find

$$\left(E - \frac{\tau}{2} \Lambda_1\right) \left(E - \frac{\tau}{2} \Lambda_2\right) \ldots \left(E - \frac{\tau}{2} \Lambda_p\right) u^{n+1/2} = u^n. \qquad (9.5.5)$$

Multiplication of Eq. (9.5.4) by $\left(E - \frac{\tau}{2} \Lambda_1\right) \ldots \left(E - \frac{\tau}{2} \Lambda_p\right)$ results in

$$\left(E - \frac{\tau}{2} \Lambda_1\right) \left(E - \frac{\tau}{2} \Lambda_2\right) \ldots \left(E - \frac{\tau}{2} \Lambda_p\right) \frac{u^{n+1} - u^n}{\tau}$$
$$= \left(E - \frac{\tau}{2} \Lambda_1\right) \ldots \left(E - \frac{\tau}{2} \Lambda_p\right) \Lambda\, u^{n+1/2} +$$
$$+ \left(E - \frac{\tau}{2} \Lambda_1\right) \ldots \left(E - \frac{\tau}{2} \Lambda_p\right) F_n. \qquad (9.5.6)$$

If operators Λ_i are commutative, then using Eq. (9.5.5) the following scheme in whole steps is derived

$$\left(E - \frac{\tau}{2} \Lambda_1\right) \ldots \left(E - \frac{\tau}{2} \Lambda_p\right) \frac{u^{n+1} - u^n}{\tau}$$
$$= \Lambda\, u^n + \left(E - \frac{\tau}{2} \Lambda_1\right) \ldots \left(E - \frac{\tau}{2} \Lambda_p\right) F_n. \qquad (9.5.7)$$

By expanding it into series with respect to τ we have

$$\frac{u^{n+1} - u^n}{\tau} = \Lambda \frac{u^n + u^{n+1}}{2} - \frac{\tau^2}{4} \Phi \left(\frac{u^{n+1} - u^n}{\tau}\right) +$$
$$+ \left(E - \frac{\tau}{2} \Lambda_1\right) \ldots \left(E - \frac{\tau}{2} \Lambda_p\right) F_n, \qquad (9.5.8)$$

$$\Phi = \sum_{i<j} \Lambda_i \Lambda_j - \frac{\tau}{2} \sum_{i<j<k} \Lambda_i \Lambda_j \Lambda_k + \cdots +$$
$$+ (-1)^{p-2} \left(\frac{\tau}{2}\right)^{p-2} \Lambda_1 \ldots \Lambda_p. \qquad (9.5.9)$$

If we assume that

$$F_n = \left(E - \frac{\tau}{2} \Lambda_p\right)^{-1} \ldots \left(E - \frac{\tau}{2} \Lambda_1\right)^{-1} f_n, \qquad (9.5.10)$$

then we arrive at the following scheme

$$\frac{u^{n+1} - u^n}{\tau} = \Lambda \frac{u^n + u^{n+1}}{2} - \frac{\tau^2}{4} \Phi \left(\frac{u^{n+1} - u^n}{\tau}\right) + f_n, \qquad (9.5.11)$$

which is of the second order of approximation in t with respect to scheme (9.5.1).

In the case when eigenvalues of operators Λ_s are negative, the stability of Eqs. (9.5.3) and (9.5.4) is proved to follow from the stability of Eq. (9.5.1) by the method described in Sec. 9.4.

9.6 The method of establishing the steady state

The majority of iterative schemes for the solution of the steady boundary value problems for the steady equation

$$L\,u - f = 0 \tag{9.6.1}$$

(L is the elliptic operator) are integration schemes of the corresponding unsteady parabolic equation

$$\frac{\partial u}{\partial t} = L\,u - f. \tag{9.6.2}$$

It is evident that the iterative process reproduces the established process of the unsteady difference solution which in fact is an approximate solution of Eq. (9.6.2). However, such a correspondence between the iterative process and the evolution in time of the solution for Eq. (9.6.2) is not necessary. Eq. (9.6.1) can be introduced into the corresponding equation of a more general structure

$$P\left(u, \frac{\partial u}{\partial t}\right) = L\,u - f, \tag{9.6.3}$$

where the operator $P[u, (\partial u/\partial t)]$ has the form of matrix which acts on the vector $[u, (\partial u/\partial t), \ldots, (\partial^q u/\partial t^q)]$. The operator $P[u, (\partial u/\partial t)]$ is called the relaxation operator and it can be of very general structure, provided Eq. (9.6.3) defines a steady regime for an arbitrary solution of Eq. (9.6.3). Schemes of such a nature were analyzed in works of V. K. Saul'ev [27], N. N. Vladimirova, B. G. Kuznetsov, and the author [74] and were given in Ch. 4, 5, and 8 of this monograph. As was indicated in Sec. 8.1, the relaxation operator $P[u, (\partial u/\partial t)]$ can also depend on the point $x_1, \ldots, x_m$ and this accelerates the rate of convergence. The relaxation operator is introduced directly into the iterative scheme. Let Λ be the approximation of operator L. Consider a universal representation of the two-layer iterative scheme (see [87])

$$B\left(\frac{u^{n+1} - u^n}{\tau}\right) = \Lambda\,u^n - f \tag{9.6.4}$$

with the indeterminate operator B which can also depend on the index n of the iteration. It is clear that scheme (9.6.4) satisfies the condition of complete consistency and is strongly stable, provided

$$\|C\| = \|E + \tau\,B^{-1}\Lambda\| \leq 1 - \varepsilon, \qquad \varepsilon > 0. \tag{9.6.5}$$

At present two methods for the construction of iterative schemes based on Eq. (9.6.4) are formulated.

The method of the majorant operator which was suggested by E. G. D'yakonov [46, 47] consists of the replacement of operator Λ by an operator B, which satisfies the condition

$$\delta_0(B\,u, u) \leq (\Lambda\,u, u) \leq \delta_1(B\,u, u), \tag{9.6.6}$$

where δ_0 and δ_1 are positive constants. Then equation

$$B\,u^{n+1} = B\,u^n + \tau(\Lambda\,u^n - f) \tag{9.6.7}$$

can be solved with respect to u^{n+1} at given u^n by iterations. In order to simplify the realization B should have a simpler structure than Λ. Thus, if

$$\Lambda = \Lambda_{11} + \Lambda_{22} + 2a_{12}\Lambda_{12},$$

then

$$B = \Lambda_{11} + \Lambda_{22}.$$

In the method of the stabilizing operator (see [51]) it is assumed for simplicity of realization that operator B is factorized

$$B = B_1 \ldots B_p, \tag{9.6.8}$$

where operators $B_1, \ldots, B_p$ can be inverted in order and operator C from Eq. (9.6.5) is strongly stable. In particular, operators $B_1, \ldots, B_p$ could have the form

$$E - \tau\,\Lambda_s$$

and in the case of non-positive operators Λ_s, they can be inverted in order. Notice that it is not absolutely necessary to satisfy the condition

$$\sum_{s=1}^{p} \Lambda_s = \Lambda.$$

The condition of strong stability (9.6.5) is the only condition. The following multi-layer iterative schemes

$$B_1\left(\frac{\Delta_0}{\tau}\right) u^n + B_2\left(\frac{\Delta_0}{\tau}\right)^2 u^n + \cdots + B_q\left(\frac{\Delta_0}{\tau}\right)^q u^n = \Lambda\,u^n + f \tag{9.6.9}$$

with indeterminate operators $B_1, \ldots, B_q$ can be considered. Scheme (9.6.9), for any operators $B_1, \ldots, B_q$ will fulfill the condition of complete consistency.

The method of the stabilizing operator can be used for the derivation of schemes of higher accuracy and also in the case of equations with variable coefficients. The first schemes of the stabilizing operator type were derived by A. A. Samarskii [63, 64] and E. G. D'yakonov [90, 95].

Chapter 10

The Method of Weak Approximation and the Construction of the Solution of the Cauchy Problem in Banach Space

10.1 Examples

Up to this point the method of fractional steps was considered as a method for the construction of economical difference schemes. Let us now show that the method of fractional steps can also be applied to differential equations. In this case the method can be considered as the method of a weak approximation of special type. We start with the simplest examples.

(i) For the Cauchy problem

soln, x(t)=t

$$\frac{dx}{dt} = f(t) \equiv 1, \qquad x(0) = 0, \tag{10.1.1}$$

the difference scheme of fractional steps is applied

$$\frac{x^{n+1/2} - x^n}{\tau} = 1, \tag{10.1.2a}$$

$$\frac{x^{n+1} - x^{n+1/2}}{\tau} = 0, \tag{10.1.2b}$$

$$x^0 = 0. \tag{10.1.2c}$$

The corresponding scheme in whole steps is

$$\frac{x^{n+1} - x^n}{\tau} = 1, \qquad x^0 = 0. \tag{10.1.3}$$

It follows from this that Eq. (10.1.2) produces an exact solution of problem (10.1.1). Scheme (10.1.2) can at the same time be interpreted as follows: at the first half step (10.1.2a) we solve the equation

$$\frac{1}{2}\,\frac{dx}{dt} = 1. \tag{10.1.4a}$$

At the second half step (10.1.2b) we solve

$$\frac{1}{2}\,\frac{dx}{dt} = 0. \tag{10.1.4b}$$

In a whole step we solve the equation

$$\frac{dx}{dt} = f(\tau, t), \qquad x^0 = 0, \tag{10.1.5}$$

where the function $f(\tau, t)$ is determined in such a way that

$$f(\tau, t) = 2, \qquad n\tau < t \leq (n + \tfrac{1}{2})\tau;$$

$$f(\tau, t) = 0, \qquad (n + \tfrac{1}{2})\tau < t \leq (n + 1)\tau.$$

Fig. 15 shows comparisons of the function $f(t) \equiv 1$, $f(\tau, t)$ and of the solutions $x(t)$ of Eq. (10.1.1) and $x(\tau, t)$ of Eq. (10.1.5).

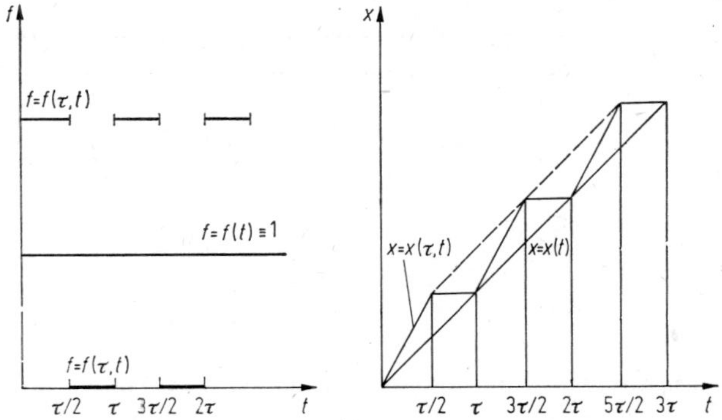

Fig. 15. Comparison of functions $f(t) \equiv 1$, $f(\tau, t)$ and of corresponding integral curves $x(t)$, $x(\tau, t)$

It is easy to see that the function $f(\tau, t)$ converges to the function $f(t) \equiv 1$ weakly, i.e.,

$$\int_{t_1}^{t_2} [f(\tau, s) - f(s)]\, ds \to 0, \qquad \tau \to 0, \qquad t_1, t_2 \text{ arbitrary.} \tag{10.1.6}$$

At the same time the corresponding solution of Eq. (10.1.5) converges strongly to the solution of Eq. (10.1.1).

(ii) For the Cauchy problem

$$\frac{dx}{dt} + a x = 0, \qquad x(0) = 1, \qquad a > 0, \tag{10.1.7}$$

we apply the difference scheme

$$\frac{x^{n+1/2} - x^n}{\tau} + a x^n = 0, \tag{10.1.8a}$$

$$\frac{x^{n+1} - x^{n+1/2}}{\tau} = 0, \tag{10.1.8b}$$

$$x^0 = 1. \tag{10.1.8c}$$

Scheme (10.1.8) can be considered as the solution of equation

$$\frac{dx}{dt} + a(\tau, t)\, x = 0, \qquad x(0) = 1, \tag{10.1.9}$$

where

$$a(\tau, t) = 2a, \quad n\tau < t \leq (n + \tfrac{1}{2})\tau; \qquad a(\tau, t) = 0, \quad (n + \tfrac{1}{2})\tau < t \leq (n + 1)\tau, \tag{10.1.10}$$

$a(\tau, t)$ approximates weakly $a(t) \equiv a$ in the sense of Eq. (10.1.6); the solution of Eq. (10.1.9) converges strongly to the solution of Eq. (10.1.7).

(iii) Consider the Cauchy problem for the transport equation

$$\frac{\partial u}{\partial t} + \frac{\partial u}{\partial x_1} + \frac{\partial u}{\partial x_2} = 0, \quad u(x_1, x_2, 0) = u_0(x_1, x_2). \tag{10.1.11}$$

The solution of this problem is

$$u(x_1, x_2, t) = u_0(x_1 - t, x_2 - t) = T_{-1}(t)\, T_{-2}(t)\, u_0(x_1, x_2), \tag{10.1.12}$$

where the displacement operators $T_{-1}(t)$, $T_{-2}(t)$ have the following meaning

$$T_{-1}(t) f(x_1, x_2) = f(x_1 - t, x_2); \qquad T_{-2}(t) f(x_1, x_2) = f(x_1, x_2 - t). \tag{10.1.13}$$

Thus, the solution operator $S(t)$ of Eq. (10.1.11) is

$$S(t) = T_{-2}(t)\, T_{-1}(t). \tag{10.1.14}$$

While preserving the function of initial data $u(x_1, x_2, 0) = u_0(x_1, x_2)$ replace Eq. (10.1.11) by the equation with "oscillating" coefficients

$$\frac{\partial u}{\partial t} + f_1(\tau, t)\frac{\partial u}{\partial x_1} + f_2(\tau, t)\frac{\partial u}{\partial x_2} = 0, \tag{10.1.15}$$

where

$$f_1(\tau, t) = 2; \quad f_2(\tau, t) = 0, \quad n\tau < t \leq (n + \tfrac{1}{2})\tau; \qquad f_1(\tau, t) = 0; \quad f_2(\tau, t) = 2, \quad (n + \tfrac{1}{2})\tau < t \leq (n + 1)\tau. \tag{10.1.16}$$

The solution of Eq. (10.1.15) in the interval from $n\tau$ to $(n + 1)\tau$ is equivalent to a successive solution of equations

$$\frac{\partial u}{\partial t} + 2\frac{\partial u}{\partial x_1} = 0, \quad n\tau < t \leq (n + \tfrac{1}{2})\tau \tag{10.1.17a}$$

and

$$\frac{\partial u}{\partial t} + 2\frac{\partial u}{\partial x_2} = 0, \quad (n + \tfrac{1}{2})\tau < t \leq (n + 1)\tau, \tag{10.1.17b}$$

respectively.

The transfer operator of Eq. (10.1.17a) is equal to

$$S_1\left(t + \frac{\tau}{2}, t\right) = S_1\left(\frac{\tau}{2}\right) = T_{-1}(\tau). \tag{10.1.18a}$$

The transfer operator of Eq. (10.1.17b) is equal to

$$S_2\left(t + \frac{\tau}{2}, t\right) = S_2\left(\frac{\tau}{2}\right) = T_{-2}(\tau). \tag{10.1.18b}$$

Consequently for the transfer operator $S_\tau(t+\tau, t) = S_\tau(\tau)$ of Eq. (10.1.15) we have

$$S_\tau(\tau) = S_2\left(\frac{\tau}{2}\right) S_1\left(\frac{\tau}{2}\right) = T_{-2}(\tau)\, T_{-1}(\tau) = S(\tau),$$

where $S(\tau)$ is the transfer operator of Eq. (10.1.11).

Eq. (10.1.15) approximates weakly Eq. (10.1.11), and the operator $S_\tau(t, 0)$ of the solution of Eq. (10.1.15) approximates strongly the solution operator $S(t, 0)$ of Eq. (10.1.11) and it coincides with the latter for $t = n\tau$.

(iv) For the equation of heat conduction

$$\frac{\partial u}{\partial t} = \frac{\partial^2 u}{\partial x_1^2} + \frac{\partial^2 u}{\partial x_2^2} \tag{10.1.19}$$

the following equation with oscillating coefficients is derived

$$\frac{\partial u}{\partial t} = a_1(\tau, t) \frac{\partial^2 u}{\partial x_1^2} + a_2(\tau, t) \frac{\partial^2 u}{\partial x_2^2}, \tag{10.1.20}$$

where

$$\begin{aligned} a_1(\tau, t) &= 2; & a_2(\tau, t) &= 0, & n\tau &< t \le (n + \tfrac{1}{2})\tau, \\ a_1(\tau, t) &= 0; & a_2(\tau, t) &= 2, & (n + \tfrac{1}{2})\tau &< t \le (n + 1)\tau. \end{aligned} \tag{10.1.21}$$

Then again we establish without any difficulty that

$$S(\tau) = S_2\left(\frac{\tau}{2}\right) S_1\left(\frac{\tau}{2}\right) = S_\tau(\tau), \tag{10.1.22}$$

where $S(\tau)$ is the transfer operator of Eq. (10.1.19); $S_i(\tau)$ is the transfer operator of equations

$$\frac{\partial u}{\partial t} = 2 \frac{\partial^2 u}{\partial x_i^2}, \qquad i = 1, 2; \tag{10.1.23}$$

$S_\tau(\tau)$ is the transfer operator of Eq. (10.1.20)[1]. Let us introduce now the idea of weak approximation for differential operators of general structure.

---

[1] An analogous interpretation of the scheme of splitting was presented in [36, 13, 79, 84, 91]. In [92] a generalization of the a.d. scheme is given for spatial differential operators. In all these papers the transformation from one fractional step to another is discrete, i.e., the authors arrive at difference schemes. Interesting splitting schemes were proposed for various problems of mathematical physics in [105, 108]. Undoubtedly, there exists close relation between splitting and the theory of semigroups [109—111].

10.2 A weak approximation for a system of differential equations

We start with the definition of weak approximation of a function.

Definition. *A set of functions $f_\tau(x, t)$ approximates weakly the function $f(x, t)$ with respect to t in the interval $[0, T]$, provided*

$$\int_{t_1}^{t_2} [f_\tau(x, s) - f(x, s)]\, ds = \delta(x, t_1, t_2, \tau) \tag{10.2.1}$$

and $\|\delta\| \to 0$ as $\tau \to 0$ for any $t_1, t_2 \in [0, T]$.

A set of linear differential operators $L_\tau(t)$ approximates weakly the operator $L(t)$ with respect to t, provided a weak approximation exists for the coefficients. It is obvious that an analogous determination of a weak approximation can be introduced for each space variable. However, we consider here only a weak approximation with respect to t and one of a special type. For brevity we speak below about operators L_τ instead of a set of operators.

Let

$$\frac{\partial u}{\partial t} = L u + f \tag{10.2.2}$$

be a system of the type considered in Ch. 1 for which the Cauchy problem is set and which is well posed in the sense of Sec. 1.1

$$u(x, 0) = u_0(x). \tag{10.2.2'}$$

Let

$$L = L_1 + L_2 + \cdots + L_p \tag{10.2.3}$$

be a representation of the operator L as a sum of operators $L_1, \ldots, L_p$ of simpler structure, generally speaking, than the operator L. Consider the operator

$$L_\tau = \alpha_1(\tau, t)\, L_1 + \alpha_2(\tau, t)\, L_2 + \cdots + \alpha_p(\tau, t)\, L_p, \tag{10.2.4}$$

where the function $\alpha_s(\tau, t)$ is determined in the following way:

$$\alpha_s(\tau, t) = p\, \delta_{s i}, \tag{10.2.5}$$

if

$$t \in \left(n\tau + \frac{i-1}{p}\tau,\ n\tau + \frac{i\tau}{p}\right]. \tag{10.2.6}$$

Here

$$\delta_{s i} = \begin{cases} 1, & i = s, \\ 0, & i \neq s, \quad i, s = 1, \ldots, p. \end{cases} \tag{10.2.7}$$

It is not difficult to show that the operator L_τ approximates weakly the operator L. Along with system (10.2.2) we also consider the system

$$\frac{\partial u}{\partial t} = L_\tau u + f_\tau, \tag{10.2.8}$$

where the function f_τ can approximate f both weakly and strongly (in particular to coincide); and the operator L_τ is determined by Eq. (10.2.4).

What should we understand by the solution $u = u_\tau$ of the Cauchy problem

$$u_\tau(x, 0) = u_0(x) \tag{10.2.8'}$$

for Eq. (10.2.8)?

Let $u_\tau(x, t)$ be the solution of the Cauchy problem for the equation

$$\frac{\partial u}{\partial t} = p\, L_1\, u$$

in the interval $0 < t \leq \tau/p$ with initial data $u(x, 0) = u_0(x)$. Determine now $u_\tau(x, t)$ in the interval $\tau/p < t \leq 2\tau/p$ as the solution of the Cauchy problem for the equation $\partial u/\partial t = p\, L_2\, u$ with initial data $u(x, \tau/p) = u_\tau(x, \tau/p)$. In an analogous way $u_\tau(x, t)$ is determined in the interval $\frac{p-1}{p}\tau < t \leq \tau$ as the solution of the Cauchy problem for the equation $\partial u/\partial t = p\, L_p\, u$ with initial data

$$u\left(x, \frac{p-1}{p}\tau\right) = u_\tau\left(x, \frac{p-1}{p}\tau\right).$$

After this the process of determination of functions u_τ is repeated. Function $u_\tau(x, t)$ constructed in this way, by calculation, is the solution of the Cauchy problem (10.2.8) and (10.2.8').

We introduce the following notation:

$S(t_2, t_1)$ is the transfer operator of Eq. (10.2.2);
$S_\tau(t_2, t_1)$ is the transfer operator of Eq. (10.2.8);
$S_i(t_2, t_1)$ is the transfer operator of the equation

$$\frac{1}{p}\frac{\partial u}{\partial t} = L_i\, u, \qquad i = 1, 2, \ldots, p.$$

Then from the construction of the solution u_τ we establish the relation

$$S_\tau(t + \tau, t) = S_p\left(t + \tau, t + \frac{p-1}{p}\tau\right) \times$$
$$\times\, S_{p-1}\left(t + \frac{p-1}{p}\tau, t + \frac{p-2}{p}\tau\right) \ldots S_1\left(t + \frac{\tau}{p}, t\right). \tag{10.2.9}$$

In conformity with the property (10.2.9) the system (10.2.8) is called the factorized or split system. Our aim is to compare properties of the operators

$$S(t_2, t_1),\ S_\tau(t_2, t_1),\ S_i(t_2, t_1).$$

Let us first introduce several definitions. We determine firstly the extended system. Let the operator L be

$$L = \sum_\alpha a_{\alpha_1 \ldots \alpha_m} D_1^{\alpha_1} \ldots D_m^{\alpha_m}. \tag{10.2.10}$$

We introduce for consideration the vector $\overset{1}{u}$ with the components

$$p^\alpha = p^{\alpha_1, \ldots, \alpha_m} = D_1^{\alpha_1} \ldots D_m^{\alpha_m} u = D^\alpha u, \tag{10.2.11}$$

where the combinations $\alpha = (\alpha_1, \ldots, \alpha_m)$ are taken in Eq. (10.2.10) and where the set $(0, \ldots, 0)$ corresponds to the initial vector $u = p^{0, \ldots, 0}$.

Apply the operator D^α to Eq. (10.2.2)

$$D^\alpha \frac{\partial u}{\partial t} = \frac{\partial}{\partial t} D^\alpha u = \frac{\partial}{\partial t} p^\alpha = D^\alpha (L u) + D^\alpha f. \tag{10.2.11'}$$

Using the differentiation rule

$$D_i (a p^\beta) = D_i a p^\beta + a D_i p^\beta,$$

where a is a certain matrix coefficient; $p^\beta = p^{\beta_1, \ldots, \beta_m}$; expression (10.2.11′) is transformed into

$$\frac{\partial p^\alpha}{\partial t} = l^\alpha \overset{1}{u} + D^\alpha f, \tag{10.2.12}$$

where l^α is some differential operator.

After combining Eqs. (10.2.12) we obtain the system

$$\frac{\partial \overset{1}{u}}{\partial t} = \overset{1}{L} \overset{1}{u} + \overset{1}{f}. \tag{10.2.13}$$

System (10.2.13) is called the first extended system which corresponds to system (10.2.2), or simply, the first extension of the system (10.2.2). An example of such a system is shown below.

Consider the equation

$$\frac{\partial u}{\partial t} + \sum_\alpha a_\alpha \frac{\partial u}{\partial x^\alpha} = 0, \quad \alpha = 1, 2. \tag{10.2.14}$$

Assuming that $p^i = \partial u / \partial x^i$ and differentiating Eq. (10.2.14) with respect to x^i we obtain

$$\frac{\partial p^i}{\partial t} + \sum_\alpha \frac{\partial a_\alpha}{\partial x^i} \frac{\partial u}{\partial x^\alpha} + \sum_\alpha a_\alpha \frac{\partial p^i}{\partial x^\alpha} = 0, \quad i, \alpha = 1, 2. \tag{10.2.15}$$

After the introduction of the vector $\overset{1}{u} = (u, p^1, p^2)$ Eqs. (10.2.14) and (10.2.15) can be represented as follows

$$\frac{\partial \overset{1}{u}}{\partial t} = \overset{1}{L} \overset{1}{u}, \tag{10.2.16}$$

$$\overset{1}{L} = - \begin{Vmatrix} \sum_\alpha a_\alpha \frac{\partial}{\partial x^\alpha} & 0 & 0 \\ \sum_\alpha \frac{\partial a_\alpha}{\partial x^1} \frac{\partial}{\partial x^\alpha} & \sum_\alpha a_\alpha \frac{\partial}{\partial x^\alpha} & 0 \\ \sum_\alpha \frac{\partial a_\alpha}{\partial x^2} \frac{\partial}{\partial x^\alpha} & 0 & \sum_\alpha a_\alpha \frac{\partial}{\partial x^\alpha} \end{Vmatrix}. \tag{10.2.17}$$

In the particular case when the coefficients are constant, any derivative p^α satisfies the same equation as function u; the operator $\overset{1}{L}$ has diagonal form

$$\overset{1}{L} = \begin{Vmatrix} L & 0 & 0 \\ 0 & L & 0 \\ 0 & 0 & L \end{Vmatrix}, \tag{10.2.18}$$

where the operator matrix (10.2.18) acts in the space of components of the vector $\overset{1}{u}$.

As before, we obtain for $\overset{2}{u}$, $\overset{3}{u}$ the second, third etc., extended systems while constructing vectors $\overset{2}{u} = \{p^{\alpha+\beta}\}$, $\overset{3}{u} = \{p^{\alpha+\beta+\gamma}\}, \ldots$, where $\beta, \gamma, \ldots$ go through the same combinations as α. The k-th extended system can be presented as

$$\frac{\partial \overset{k}{u}}{\partial t} = \overset{k}{L}\,\overset{k}{u} + \overset{k}{f}. \tag{10.2.19}$$

It is easy to establish that the following representation corresponds to Eq. (10.2.3)

$$\overset{k}{L} = \overset{k}{L}_1 + \cdots + \overset{k}{L}_p \tag{10.2.20}$$

and the operator

$$\overset{k}{L}_\tau = \alpha_1 \overset{k}{L}_1 + \cdots + \alpha_p \overset{k}{L}_p \tag{10.2.21}$$

corresponds to operator L_τ. Thus, along with system (10.2.8) the following extended system is considered

$$\frac{\partial \overset{k}{u}}{\partial t} = \overset{k}{L}_\tau \overset{k}{u} + \overset{k}{f}_\tau. \tag{10.2.22}$$

We define the symbols:

$\overset{k}{S}(t_2, t_1)$ is the transfer operator of Eq. (10.2.19);

$\overset{k}{S}_\tau(t_2, t_1)$ is the transfer operator of Eq. (10.2.22);

$\overset{k}{S}_i(t_2, t_1)$ is the transfer operator of the equation $\frac{1}{p}\frac{\partial \overset{k}{u}}{\partial t} = \overset{k}{L}_i \overset{k}{u}$.

The following relation can be established

$$\overset{k}{S}_\tau(t+\tau, t) = \overset{k}{S}_p\left(t+\tau, t+\frac{p-1}{p}\tau\right) \times$$
$$\times \overset{k}{S}_{p-1}\left(t+\frac{p-1}{p}\tau, t+\frac{p-2}{p}\tau\right) \ldots \overset{k}{S}_1\left(t+\frac{\tau}{p}, \tau\right). \tag{10.2.23}$$

Eq. (10.2.23) can be formulated in the following manner: the factorization of an extended system is a factorized extension. In other words, operations of extension and factorization are commutative.

The extended Cauchy problem corresponds to an extended system. For this it is sufficient to assume that

$$p_0^\alpha = D^\alpha u_0, \quad p_0^{\alpha+\beta} = D^{\alpha+\beta} u_0 \tag{10.2.24}$$

etc. The Cauchy problem for systems (10.2.2) and (10.2.8) and their extensions are denoted by

$$I, I_\tau, \overset{k}{I}, \overset{k}{I}_\tau.$$

It is also assumed that problems I and I_τ are determined in the same Banach space B, problems $\overset{k}{I}$ and $\overset{k}{I}_\tau$ in corresponding Banach spaces B_k. For example, a norm in B_k can be determined in the following way:

$$\|\overset{k}{u}\|_{B_k}^2 = \sum_{\alpha_1, \ldots, \alpha_k} \|p^{\alpha_1+\alpha_2+\cdots+\alpha_k}\|_B^2.$$

Definitions of several properties of solutions of uniformly correct systems are given below. (In Lemma 1 and 2 we consider homogeneous problems.)

Lemma 1. *If the system* (10.2.2) *is uniformly correct, then the solution* $u(t)$ *is uniformly continuous.*

Proof. Because of the continuity of operator $S(t+\tau, t)$, the function $u(t)$ is continuous with respect to t for any t within the interval $[0, T]$. According to Cantor's theorem $u(t)$ is uniformly continuous. □

Lemma 2. *If* (a) *problems I and $\overset{1}{I}$ are uniformly correct and* (b) *coefficients a_α are uniformly continuous with respect to t, then expressions $\partial u/\partial t$ and $L u$ exist and are uniformly continuous with respect to t, satisfying the equation*

$$\frac{\partial u}{\partial t} = L u.$$

Proof. The vector $\overset{1}{u}(t) = \{p^\alpha(t)\}$ exists for arbitrary $t \in [0, T]$. Consequently, expression $q(t) = L u = \sum_\alpha a_\alpha(t) p^\alpha(t)$ also exists for $t \in [0, T]$. According to Lemma 1, $p^\alpha(t)$ is uniformly continuous as the component of the uniformly continuous vector $\overset{1}{u}(t)$. Considering the condition of Lemma 2 we see that $L u$ is uniformly continuous.

From the definition of the generalized solution we have

$$u(t) = \lim u(\varepsilon, t), \quad \varepsilon \to 0,$$

where the limit is uniform with respect to t in B_1, and $u(\varepsilon, t)$ is the analytic solution of I which belongs to B_1. They satisfy

$$\frac{\partial u(\varepsilon, t)}{\partial t} = L(t)\, u(\varepsilon, t), \tag{10.2.25 a}$$

$$u(\varepsilon, t+\tau) - u(\varepsilon, t) = \int^{t+\tau} L(\theta)\, u(\varepsilon, \theta)\, d\theta. \tag{10.2.25 b}$$

Proceeding to the limit in Eq. (10.2.25b) with the norm $B_0 = B$ we obtain

$$u(t+\tau) - u(t) = \int_t^{t+\tau} L(\theta)\, u(\theta)\, d\theta. \tag{10.2.26}$$

Since $L(t)\, u(t)$ is uniformly continuous, then

$$\frac{u(t+\tau) - u(t)}{\tau} \to \frac{\partial u(t)}{\partial t} = L(t)\, u(t)$$

and $\partial u/\partial t$ is uniformly continuous. The lemma is proved. □

Lamma 3. *If* (a) *I and $\overset{1}{I}$ are uniformly correct*; (b) $u_0 \in B_1$, *then $f(t) \in B_1$ and is uniformly continuous with respect to t, then formula*

$$u(t) = S(t, 0)\, u_0 + \int_0^t S(t, \theta)\, f(\theta)\, d\theta \tag{10.2.27}$$

produces the solution

$$\frac{\partial u}{\partial t} = L u + f \tag{10.2.28}$$

with initial data

$$u(0) = u_0. \tag{10.2.29}$$

Proof. Examine first the properties of functions

$$F(t, \theta) = S(t, \theta)\, f(\theta). \tag{10.2.30}$$

For fixed θ, $F(t, \theta)$ is the solution $u(t)$ of system I

$$\frac{\partial u}{\partial t} = L u$$

with initial data $u(\theta) = f(\theta)$. According to Lemma 1, $F(t, \theta)$ is uniformly continuous with respect to t in B and B_1. Let us evaluate in the norm of B in the difference

$$\begin{aligned} F(t, \theta + h) - F(t, \theta) &= S(t, \theta + h)\, f(\theta + h) - S(t, \theta)\, f(\theta) \\ &= S(t, \theta + h)\, [f(\theta + h) - f(\theta)] + \\ &\quad + [S(t, \theta + h) - S(t, \theta)]\, f(\theta) \\ &= S(t, \theta + h)\, [f(\theta + h) - f(\theta)] + \\ &\quad + S(t, \theta + h)\, [E - S(\theta + h, \theta)]\, f(\theta). \end{aligned} \tag{10.2.31}$$

The first term in Eq. (10.2.31) approaches zero as $h \to 0$ uniformly with respect to θ because of uniform continuity in $f(\theta)$. Evaluate now the second term. There we have

$$[E - S(\theta + h, \theta)]\, f(\theta) = -[F(\theta + h, \theta) - f(\theta)].$$

According to the properties of $F(t, \theta)$ this term approaches zero as $h \to 0$ for any θ.

Consequently, the function $F(t, \theta)$ is continuous with respect to θ and this means that it is uniformly continuous within the segment $[0, t]$, and the integral $F(t) = \int_0^t F(t, \theta)\, d\theta$ exists in B. It is shown analogously that it exists in B_1 also.

Consider the difference

$$u(t+\tau) - u(t) = [S(t+\tau, 0) - S(t, 0)]\, u_0 + $$
$$+ \int_t^{t+\tau} S(t+\tau, \theta)\, f(\theta)\, d\theta + \int_0^t [S(t+\tau, \theta) - S(t, \theta)]\, f(\theta)\, f(\theta)\, d\theta$$
$$= [S(t+\tau, t) - E]\, S(t, 0)\, u_0 + \int_t^{t+\tau} S(t+\tau, \theta)\, f(\theta)\, d\theta + $$
$$+ \int_0^t [S(t+\tau, t) - E]\, S(t, \theta)\, f(\theta)\, d\theta. \tag{10.2.32}$$

Using the expression (10.2.27) we can transform Eq. (10.2.32) into

$$u(t+\tau) - u(t) = [S(t+\tau, t) - E]\, [u(t) - \int_0^t S(t, \theta)\, f(\theta)\, d\theta] + $$
$$+ \int_t^{t+\tau} S(t+\tau, \theta)\, f(\theta)\, d\theta + \int_0^t [S(t+\tau, t) - E]\, S(t, \theta)\, f(\theta)\, d\theta$$
$$= [S(t+\tau, t) - E]\, u(t) + \int_t^{t+\tau} S(t+\tau, \theta)\, f(\theta)\, d\theta + $$
$$+ \int_0^t [S(t+\tau, t) - E]\, S(t, \theta)\, f(\theta)\, d\theta - $$
$$- [S(t+\tau, t) - E] \int_0^t S(t, \theta)\, f(\theta)\, d\theta. \tag{10.2.33}$$

The last two terms in Eq. (10.2.33) can be cancelled and division by τ results in

$$\frac{u(t+\tau) - u(t)}{\tau} = \frac{[S(t+\tau, t) - E]}{\tau}\, u(t) + \frac{1}{\tau} \int_t^{t+\tau} S(t+\tau, \theta)\, f(\theta)\, d\theta. \tag{10.2.34}$$

According to Lemma 3 $F(t, 0) \in B_1$; $F(t) \in B_1$ and consequently

$$u(t) = S(t, 0)\, u(0) + F(t) \in B_1,$$

because

$$u(0) \in B_1.$$

But then

$$\frac{[S(t+\tau,t)-E]}{\tau}\,u(t)=\frac{1}{\tau}\int\limits_{t}^{t+\tau}L(\theta)\,U(\theta)\,d\theta, \qquad (10.2.35)$$

where $U(\theta)$ is the solution of the Cauchy problem

$$\frac{\partial U(\theta)}{\partial\theta}=L(\theta)\,U(\theta), \qquad U(t)=u(t), \qquad t\le\theta\le t+\tau.$$

Since $L(\theta)\,U(\theta)$, $S(t+\tau,\theta)\,f(\theta)$ are uniformly continuous, then using the mean value thecrem and proceeding to the limit in Eq. (10.2.34) as $\tau\to 0$ we obtain

$$\frac{\partial u(t)}{\partial t}=L(t)\,u(t)+f(t).$$

The initial data are satisfied because of Eq. (10.2.27) and the lemma is proved. □

Formula (10.2.27) has a meaning even then when $u(\theta)\in B$, $f(\theta)\in B$ and the problem I is uniformly correct. In this case it can be used as a basis for the determination of the generalized solution of problems (10.2.28) and (10.2.29).

From Lemma 3 we obtain

Corollary. *If the problem I is correct, then the solution of problem* (10.2.28), (10.2.29) *depends continuously on the right-hand side of f (correct with respect to the right-hand side). Verification starts from the relation*

$$\|u(t)\|\le\|S(t,0)\|\,\|u_0\|+\int\limits_0^t\|S(t,\theta)\|\,\|f(\theta)\|\,d\theta$$

$$\le e^{\varkappa t}\,\|u_0\|+\int\limits_0^t e^{\varkappa(t-\theta)}\,\|f(\theta)\|\,d\theta.$$

Below we take formula (10.2.27) *as the solution of Eqs.* (10.2.28), (10.2.29) *in the sense of Lemma* 3.

Lemmas 1, 2 and 3 are valid also for system I_τ.

10.3 Convergence theorems[1]

The following theorem can be proved:

Theorem 1. *If* (a) *I_τ and $\overset{1}{I}_\tau$ are uniformly correct in B and B_1;* (b) *I_τ is correct for the right-hand side;* (c) *$L_i u$ is uniformly continuous with respect to t for the solution $u(t)$ of problem I;* (d) *$u_0\in B_1$, then $u(t)$ is the only solution of I which satisfies conditions* (c), (d); *and $u_\tau(t)$ converges strongly to $u(t)$ at $\tau\to 0$ uniformly with respect to t.*

[1] This section is based on works by G. V. Demidov and the author [88, 89].

In other words, if the factorized initial and first extended systems are correct, and the solution of the initial system (10.2.2) is sufficiently smooth, then this is the only solution and the solution of the factorized system converges to it.

Proof. Let $u(t)$ be the solution of the problem I with initial data $u_0 \in B_1$, $u_\tau(t)$ is the solution of problem I_τ with the identical initial data u_0. The function

$$v(t) = u_\tau(t) - u(t) \tag{10.3.1}$$

is the solution of the problem

$$\frac{\partial v}{\partial t} = L_\tau v + (L_\tau - L)\, u(t), \qquad v(0) = 0. \tag{10.3.2}$$

Due to Lemma 3 $v(t)$ is calculated from

$$v(t) = \int_0^t S_\tau(t, \theta)\,(L_\tau - L)\, u(\theta)\, d\theta = \sum_{i=1}^{p} v_i(t), \tag{10.3.3}$$

where

$$\begin{gathered} v_i(t) = \int_0^t S_\tau(t, \theta)\, \varepsilon_i(\tau, \theta)\, \varphi_i(\theta)\, d\theta; \\ \varepsilon_i(\tau, \theta) = \alpha_i(\tau, \theta) - 1; \qquad \varphi_i(\theta) = L_i(\theta)\, u(\theta). \end{gathered} \tag{10.3.4}$$

Let $t = n\tau + \frac{i-1}{p}\tau + \eta$, $0 \le \eta \le \frac{\tau}{p}$. Then

$$\int_0^t = \int_0^{n\tau} + \int_{n\tau}^{n\tau + \frac{i-1}{p}\tau + \eta}.$$

Because of the uniform correctness of the problem I_τ the second integral is of order $O(\tau)$ uniformly with respect to t. Therefore it is sufficient to evaluate integral $v_i(t)$ at $t = n\tau$. For v_i we have the expression

$$\begin{aligned} v_i(t) &= \sum_{k=0}^{n-1} \int_{k\tau}^{(k+1)\tau} S_\tau(t, \theta)\, \varepsilon_i(\tau, \theta)\, \varphi_i(\theta)\, d\theta \\ &= \sum_{k=0}^{n-1} \sum_{j=1}^{p} \int_{\left(k + \frac{j-1}{p}\right)\tau}^{(k + j/p)\tau} S_\tau(t, \theta)\, \varepsilon_i(\tau, \theta)\, \varphi_i(\theta)\, d\theta. \end{aligned} \tag{10.3.5}$$

Taking into account Eqs. (10.2.5) and (10.3.4), the following expression is valid for ε_i:

$$\varepsilon_i(\tau, \theta) = p\,\delta_{ij} - 1, \quad \theta \in \left[\left(k + \frac{j-1}{p}\right)\tau, \left(k + \frac{j}{p}\right)\tau\right]. \tag{10.3.6}$$

From this for $v_i(t)$ we have

$$v_i(t) = \sum_{k=0}^{n-1} \sum_{j=1}^{p} \int_{\left(k+\frac{j-1}{p}\right)\tau}^{(k+j/p)\tau} \left[S_\tau\left(t, \theta + \frac{i-j}{p}\tau\right) \varphi_i\left(\theta + \frac{i-j}{p}\tau\right) - \right.$$

$$\left. - S_\tau(t, \theta)\, \varphi_i(\theta)\right] d\theta. \tag{10.3.7}$$

We see that the evaluation of $v_i(t)$ is reduced to the evaluation of an expression of the type

$$\int_a^{a+h_1} [S_\tau(t, \theta + h_2)\, \varphi(\theta + h_2) - S_\tau(t, \theta)\, \varphi(\theta)]\, d\theta \tag{10.3.8}$$

for a uniformly correct operator $S_\tau(t, \theta)$ and for a uniformly continuous function $\varphi(\theta)$ with respect to θ.

Expression (10.3.8) can be represented as the sum

$$\int_a^{a+h_1} [S_\tau(t, \theta + h_2) - S_\tau(t, \theta)]\, \varphi(\theta)\, d\theta +$$

$$+ \int_a^{a+h_1} S_\tau(t, \theta + h_2)\, [\varphi(\theta + h_2) - \varphi(\theta)]\, d\theta. \tag{10.3.9}$$

Because of the uniform correctness of $S_\tau(t, \theta)$ and of uniform continuity of $\varphi(\theta)$ the second term of Eq. (10.3.9) is of order $h_1 \alpha(h_2)$ where $\alpha(h_2) \to 0$ as $h_2 \to 0$ and the expression

$$I_2 \varphi_i = \sum_{k=0}^{n-1} \sum_{j=1}^{p} \int_{\left(k+\frac{j-1}{p}\right)\tau}^{(k+j/p)\tau} S_\tau\left(t, \theta + \frac{i-j}{p}\tau\right) \left[\varphi_i\left(\theta + \frac{i-j}{p}\tau\right) - \varphi_i(\theta)\right] d\theta$$

is of order $\alpha(\tau)$ with respect to the norm.

Let us evaluate the first term of Eq. (10.3.9). Assume, for simplicity, that h_2 is positive. Then using the relation

$$S_\tau(t, \theta) = S_\tau(t, \theta + h_2)\, S_\tau(\theta + h_2, \theta), \tag{10.3.10}$$

expression (10.3.9) can be transformed into

$$\int_a^{a+h_1} S_\tau(t, \theta + h_2)\, [E - S_\tau(\theta + h_2, \theta)]\, \varphi(\theta)\, d\theta. \tag{10.3.11}$$

Consider the expression

$$[S_\tau(\theta + h_2, \theta) - E]\, \varphi(\theta). \tag{10.3.12}$$

Let $U(t)$ be the solution of the Cauchy problem

$$\frac{\partial U}{\partial t} = L_\tau U, \qquad U(\theta) = \varphi(\theta).$$

Then Eq. (10.3.12) is no other than $U(\theta + h_2) - U(\theta)$. If $\varphi(\theta) \in B_1$, then we have

$$U(\theta + h_2) - U(\theta) = \int_\theta^{\theta + h_2} L_\tau U\, dt.$$

Because of the correctness of I_τ, $L_\tau U$ exists, is bounded and $\int_\theta^{\theta+h_2} L_\tau \varphi \, d\theta = O(h_2)$ uniformly with respect to t, τ.

Thus, the integral

$$\int_a^{a+h_1} [S_\tau(t, \theta + h_2) - S_\tau(t, \theta)] \, \varphi(\theta) \, d\theta$$

for $\varphi(\theta) \in B_1$ is also of order $O(h_1 h_2)$ and consequently for $\varphi(\theta) \in B_1$ the expression

$$I_1 \varphi_i = \sum_{k=0}^{n-1} \sum_{j=1}^{p} \int_{\left(k+\frac{j-1}{p}\right)\tau}^{(k+j/p)\tau} \left[S_\tau\left(t, \theta + \frac{i-j}{p}\tau\right) - S_\tau(t, \theta)\right] \varphi_i(\theta) \, d\theta \tag{10.3.13}$$

is of order $O(\tau)$ with respect to the norm. I_1 can be considered as a set of operators $I_1(\tau)$ which operate on the function $u(t)$. Since

$$\| I_1(\tau) \varphi_i \| \leq C(T) \max \| L_i u \|, \tag{10.3.14}$$

then for $u \in B_1$ and $\varphi_i \in B_0$ the set of $I_1(\tau)$ is uniformly bounded. Since $u \in B_2$, $\| I_1(\tau) \varphi \| \to 0$, then according to the theorem of Banach-Steinhaus $I_1(\tau)$ converges to the null operator, i.e., for $\varphi_i = L_i u \in B_0$ the expression $I_1(\tau) \to 0$ as $\tau \to 0$.

Noting that $v = \sum_{i=1}^{p} (I_1 + I_2) \varphi_i$ we arrive at the final estimate

$$\| v(t) \| \to 0 \tag{10.3.15}$$

as $\tau \to 0$. Theorem 1 is thus proved. □

Notice that for equations with constant coefficients it is sufficient that problem I_τ be correct, and the conditions for Theorem 1 reduce to those of sufficient smoothness of the initial data. This is valid for all equations in which the correctness of the extended system follows from the correctness of the initial system.

Theorem 1 proves strong convergence of the solution $u_\tau(t)$ of the factorized system (10.2.8), with oscillating coefficients that weakly approximate the system (10.2.2), to the solution $u(t)$ of system (10.2.2). In addition, we can deduce the existence of the solution and correctness of system (10.2.2) from the existence f the solution of the factorized system (10.2.8) and give a constructive definition of the transfer operator $S(t_2, t_1)$ by means of the operator $S_\tau(t_2, t_1)$. This statement is valid because operator $S_\tau(t_2, t_1)$ is represented as the product (10.2.9) of operators of much simpler structure.

Thus, the integration of Eq. (10.2.2) is reduced to the integration of a system of much simpler structure.

We can prove the following theorem:

Theorem 2. *If problems I_τ, $\overset{1}{I}_\tau$ and $\overset{2}{I}_\tau$ are uniformly correct, then* (a) *$u_\tau(t)$ converges uniformly to function $u(t) = S(t, 0) u_0$ with respect to t as $\tau \to 0$;* (b) *operator $S(t_2, t_1)$ is uniformly correct.*

Proof. According to the conditions of the theorem we have

$$\|\overset{k}{S_\tau}(t_2, t_1)\| \leq e^{\alpha(t_2-t_1)}, \quad 0 \leq t_1 \leq t_2 \leq T, \quad k = 0, 1, 2, \tag{10.3.16}$$

where the constant α does not depend on τ, t_1 and t_2.

Let $u_{\tau_1}(t)$, $u_{\tau_2}(t)$ be the solutions of problem I_τ which correspond to $\tau = \tau_1$ and $\tau = \tau_2$ as $u_0 \in B_2$. The function

$$v(t) = u_{\tau_2}(t) - u_{\tau_1}(t) \tag{10.3.17}$$

is the solution of the problem

$$\frac{\partial v}{\partial t} = L_{\tau_1} v + (L_{\tau_2} - L_{\tau_1}) u_{\tau_2}, \quad 0 \leq t \leq T, \tag{10.3.18}$$
$$v(0) = 0.$$

For $v(t)$ we have

$$v(t) = \int_0^t S_{\tau_1}(t, \theta)(L_{\tau_2} - L_{\tau_1}) u_{\tau_2}(\theta)\, d\theta. \tag{10.3.19}$$

$v(t)$ can be represented as

$$\left.\begin{aligned} v(t) &= I_1 + I_2; \\ I_1 &= \int_0^t S_{\tau_1}(t, \theta)[L - L_{\tau_1}] u_{\tau_2}(\theta)\, d\theta; \\ I_2 &= \int_0^t S_{\tau_1}(t, \theta)[L_{\tau_2} - L] u_{\tau_2}(\theta)\, d\theta. \end{aligned}\right\} \tag{10.3.20}$$

Evaluate now the norm of I_2 (the norm of I_1 is evaluated analogously)

$$\begin{aligned} I_2 &= \sum_{i=1}^{p} \int_0^t S_{\tau_1}(t, \theta)\, \varepsilon_i\, \varphi_i(\theta)\, d\theta; \\ \varepsilon_i &= \alpha_i(\tau_2, t) - 1; \quad \varphi_i = L_i u_{\tau_2}. \end{aligned} \tag{10.3.21}$$

As in Theorem 1 we can assume that $t = n\tau_2$ because the neglected part is of order $O(\tau_2)$. When I_2 is transformed as was done in Theorem 1 we have

$$I_2 = \sum_{i=1}^{p} \sum_{j=1}^{p} \sum_{k=0}^{n-1} \int_{\left(k+\frac{j-1}{p}\right)\tau_2}^{(k+j/p)\tau_2} \left[S_{\tau_1}\left(t, \theta + \frac{i-j}{p}\tau_2\right) \times \right.$$
$$\left. \times\, \varphi_i\left(\theta + \frac{i-j}{p}\tau_2\right) - S_{\tau_1}(t, \theta)\, \varphi_i(\theta)\right] d\theta. \tag{10.3.22}$$

Considering the fact that $\varphi_i(\theta) \in B_1$ and constructing a bound as in Theorem 1 we obtain

$$\|I_1\| = \alpha(\tau_1), \quad \|I_2\| = \alpha(\tau_2), \quad \|v(t)\| \leq \alpha(\tau_1) + \alpha(\tau_2). \tag{10.3.23}$$

This means that the sequence $u_\tau(t)$ is fundamental and converges strongly to $u(t) \in B$ as $\tau \to 0$; at the same time the convergence is uniform with respect to t. In other words, the sequence of functions $S_\tau(t, 0)\, u(0)$ converges for $u(0) \in B_2$. Because B_2 is dense in B and S_τ is bounded everywhere the conditions of the Banach-Steinhaus theorem are fulfilled. Because of this theorem a bounded operator exists (see [72])

$$S(t_2, t_1) = \lim_{\tau \to 0} S_\tau(t_2, t_1).$$

Due to the uniform correctness of operators $S_\tau(t_2, t_1)$ the operator $S(t_2, t_1)$ satisfies the uniform correctness condition

$$\| S(t_2, t_1) \| \le e^{\alpha(t_2 - t_1)}$$

with the same exponent α.

Let us show that the convolution condition (semi-group convolution) is satisfied

$$S(t_3, t_1) = S(t_3, t_2)\, S(t_2, t_1). \tag{10.3.24}$$

We have

$$S(t_3, t_2)\, S(t_2, t_1) - S(t_3, t_1) = [S_\tau(t_3, t_1) - S(t_3, t_1)] + \\ + [S(t_3, t_2) - S_\tau(t_3, t_2)]\, S_\tau(t_2, t_1) + S(t_3, t_2)\, [S(t_2, t_1) - S_\tau(t_2, t_1)]. \tag{10.3.25}$$

Proceeding to the limit in Eq. (10.3.25) as $\tau \to 0$ Eq. (10.3.24) is obtained, and this is precisely what we wanted to prove. Finally, let us prove that the condition of a continuous joining is also satisfied. We have

$$\| u(t) - u(0) \| \le \| u(t) - u_\tau(t) \| + \| u_\tau(t) - u(0) \|. \tag{10.3.26}$$

Because $\| u(t) - u_\tau(t) \| \to 0$ is uniform with respect to t as $\tau \to 0$, then τ_0 is selected small enough to satisfy

$$\| u(t) - u_\tau(t) \| < \frac{\varepsilon}{2} \tag{10.3.27}$$

for $\tau \le \tau_0$ for all t.

For fixed τ it is necessary to select t_0 to be sufficiently small that

$$\| u_\tau(t) - u(0) \| < \frac{\varepsilon}{2} \tag{10.3.28}$$

for $t \le t_0$. It follows from Eqs. (10.3.26), (10.3.27) and (10.3.28) that

$$\| u(t) - u(0) \| < \varepsilon \tag{10.3.29}$$

for $t \le t_0$. Then Theorem 2 is proved. □

Theorem 3. *If problems* $\overset{k}{I}_\tau$ $(k = 0, 1, 2, 3)$ *are well posed, then for an arbitrary* $u(0) \in B_1$, $\overset{1}{u}_\tau(t) \to \overset{1}{u}(t)$ *and the limit function* $u(t)$ *satisfies equation* $\partial u / \partial t = L u$ *and also the initial data.*

Proof. In the consideration of $\overset{1}{I}$ in B_1 we proceed identically as in the case of Theorem 2 and prove that $\overset{1}{u}_\tau$ converges to $\overset{1}{u}$. It will also be shown that $u(t)$ has a derivative $\partial u / \partial t$ and satisfies the equation

$\partial u/\partial t = L\,u$. Consider the averaged function

$$\bar{u}_\tau(t) = \frac{1}{\tau}\int\limits_t^{t+\tau} u_\tau(\theta)\,d\theta. \tag{10.3.30}$$

The function $\bar{u}_\tau(t)$ converges to $u(t)$ uniformly with respect to t as $\tau \to 0$. In fact,

$$\begin{aligned} u(t) - \bar{u}_\tau(t) &= \frac{1}{\tau}\int\limits_t^{t+\tau} [u(t) - u_\tau(\theta)]\,d\theta \\ &= \frac{1}{\tau}\int\limits_t^{t+\tau} [u(t) - u(\theta)]\,d\theta + \frac{1}{\tau}\int\limits_t^{t+\tau} [u(\theta) - u_\tau(\theta)]\,d\theta. \end{aligned} \tag{10.3.31}$$

The function $u(t)$ is uniformly continuous within the interval $0 \le t \le T$ because it is a uniform limit of the uniformly continuous functions $u_\tau(t)$. Consequently, the first term in Eq. (10.3.31) approaches zero as $\tau \to 0$ uniformly with respect to t; the second term approaches zero with respect to t because of the uniform convergence of $u_\tau(t)$ to $u(t)$. Thus,

$$u(t) - \bar{u}_\tau(t) \to 0 \tag{10.3.32}$$

as $\tau \to 0$ uniformly with respect to t. In a similar way it is proved that

$$\left\| \overset{1}{\bar{u}}(t) - \overset{1}{u}(t) \right\| = \left\| \overset{1}{u} - \overset{1}{u}(t) \right\| \to 0 \tag{10.3.33}$$

as $\tau \to 0$ uniformly with respect to t. Thus, $\bar{u}_\tau(t)$ together with derivatives $D^\alpha \bar{u}_\tau$ converge to $u(t)$. Application of averaging to equation I_τ results in

$$\frac{\partial u_\tau}{\partial t} = L\,\bar{u}_\tau + f_\tau(t), \tag{10.3.34}$$

where

$$f_\tau(t) = \frac{1}{\tau}\int\limits_t^{t+\tau} [L_\tau(\theta)\,u_\tau(\theta) - L(t)\,\bar{u}_\tau(t)]\,d\theta. \tag{10.3.35}$$

Let us show that $f_\tau(t) \to 0$ as $\tau \to 0$ for $u_0 \in B_2$. For $f_\tau(t)$ we have

$$\begin{aligned} f_\tau(t) &= \sum_{i=1}^{p} \frac{1}{\tau}\int\limits_t^{t+\tau} [\alpha_i(\tau,\theta)\,L_i(\theta)\,u_\tau(\theta) - L_i\,\bar{u}_\tau(t)]\,d\theta \\ &= \sum_{i=1}^{p} \frac{p}{\tau}\int\limits_{\sigma_i} [L_i(\theta)\,u_\tau(\theta) - L_i(t)\,\bar{u}_\tau(t)]\,d\theta, \end{aligned} \tag{10.3.36}$$

where the integration in the last integral takes place in the subinterval of length τ/p where $\alpha_i \neq 0$. This integral can be transformed as follows

$$\begin{aligned} &\int\limits_{\sigma_i} [L_i(\theta)\,u_\tau(\theta) - L_i(t)\,\bar{u}_\tau(t)]\,d\theta \\ &= \int\limits_{\sigma_i} [L_i(\theta)\,u_\tau(\theta) - L_i(t)\,u_\tau(t)]\,d\theta + \int\limits_{\sigma_i} L_i(t)\,[u_\tau(t) - \bar{u}_\tau(t)]\,d\theta. \end{aligned} \tag{10.3.37}$$

$L_i(\theta)\, u_\tau(\theta)$ is uniformly continuous and therefore the first integral with respect to the norm is of order $\tau\, \varepsilon_1(\tau)$ where $\varepsilon_1(\tau) \to 0$, $\tau \to 0$. Since $L_i(t)\, \bar{u}_\tau(t)$ converges uniformly to $L_i(t)\, u_\tau(t)$ as $\tau \to 0$, then the second integral is also of order $\tau\, \varepsilon_2(\tau)$ with respect to the norm where $\varepsilon_2(t) \to 0$, $\tau \to 0$. As a result $f_\tau(t) \to 0$, $\tau \to 0$ for $u_0 \in B_2$. It is clear that $f_\tau(t)$ can be represented as

$$f_\tau(t) = M_\tau(t)\, u_0, \tag{10.3.38}$$

where the linear operator $M_\tau(t)$ has a bounded norm in B_1 uniform with respect to τ:

$$\| M_\tau(t) \|_{B_1} \leq C(T)\, \| u_0 \|_{B_2}. \tag{10.3.39}$$

Since $\| M_\tau(t)\, u_0 \| \to 0$ and $\tau \to 0$ for all $u_0 \in B_2$, then according to the Banach-Steinhaus theorem $M_\tau(t)$ in space B_1 approaches a null operator and $f_\tau(t) \to 0$ for $u_0 \in B_1$. Due to the proved fact that $L\, \bar{u}_\tau \to L\, u$, then from Eq. (10.3.34) it follows that $\partial u_\tau/\partial t \to \partial u/\partial t$ and we have the equality $\partial u/\partial t = L\, u$. In this way Theorem 3 is proved. □

It has already been mentioned that in the case of equations with constant coefficients the correctness of $\overset{k}{I}$ $(k = 1, 2, 3)$ follows from the correctness of I. The same is valid for systems of $\overset{k}{I}_\tau$. In this case the conditions of Theorems 1–3 contain only the smoothness requirements of the initial data.

Weaker correctness requirements of systems I_τ are obtained in the case when a study is carried out for particular Banach spaces or for particular systems. In this case two theorems are valid.

Theorem 4. *If problems I_τ and $\overset{1}{I}_\tau$ are uniformly correct in $B = L_q(\Omega)$ $(q > 1)$ and B_1, then $u_\tau(t)$ as $\tau \to 0$ is fundamentally uniform with respect to t and each regular limit function $u(t)$ represents the solution of the problem I. If $\overset{1}{u}_0 \in L_q$ then $u(t)$ is regular and is the only solution of problem I.*

Notice that it is also assumed in Theorem 4 that $\overset{1}{u}$ contains at least first derivatives with respect to all space derivatives.

Consider a symmetric system of the first order

$$\frac{\partial u}{\partial t} = \sum_{i=1}^{m} A_i \frac{\partial u}{\partial x_i},$$

where $A_i(x, t)$ represents the symmetric matrices which are continuous in Ω together with first derivatives with respect to space variables. Assume that

$$L_i = A_i \frac{\partial}{\partial x_i}.$$

In this case $p = m$ and $B = L_2$. We can also prove

Theorem 5. *Problems I and I_τ be uniformly correct. The function $u_\tau(t)$ converges uniformly with respect to t as $\tau \to 0$ to the solution of problem I.*

References

1. Lyusternik, L. A., Sobolev, V. I.: Elements of functional analysis (Russian). Moscow 1965.
2. Hille, E., Phillips, R. S.: Functional analysis and semi-groups. Amer. Math. Soc. 1957.
3. Lax, P. D., Richtmyer, R. D.: Survey of the stability of linear finite difference equations. Comm. Pure Appl. Math. **9**, 267—293 (1956).
4. Richtmyer, R. D.: Difference methods for initial-value problems. Interscience, 1958.
5. Ryaben'kii, V. S.: Application of the method of finite differences to the solution of the Cauchy problem (Russian). Dokl. Akad. Nauk SSSR **86**, no. 6, 1071—1074 (1952).
6. Meiman, N. N.: The theory of equations in partial derivatives (Russian). Dokl. Akad. Nauk SSSR **97**, no. 4, 593—596 (1954).
7. Richtmyer, R. D.: Abstract theory of the linear inhomogeneous Cauchy problem (Russian). Some Probl. of Appl. and Comp. Math., Collection of Works. Novosibirsk 1966.
8. Lokutsievskii, O. V.: Numerical methods for the solution of equations in partial derivatives (Russian). Usp. Mat. Nauk **11**, no. 3 (1956).
9. Marchuk, G. I.: Computing methods for nuclear reactors (Russian). Gosatomizdat, 1961.
10. Peaceman, D. W., Rachford jr., H. H.: The numerical solution of parabolic and elliptic differential equations. J. Soc. Ind. Appl. Math. **3**, 28—41 (1955).
11. Douglas jr., J.: On the numerical integration of $u_{xx} + u_{yy} = u_t$ by implicit methods. J. Soc. Ind. Appl. Math. **3**, 42—65 (1955).
12. Douglas jr., J., Rachford jr., H. H.: On the numerical solution of heat conduction problems in two and three space variables. Trans. Amer. Math. Soc. **82**, 421—439 (1956).
13. Yanenko, N. N.: On the difference method for the calculation of multidimensional heat conduction equations (Russian). Dokl. Akad. Nauk SSSR **125**, no. 6, 1207—1210 (1959).
14. Yanenko, N. N., Suchkov, V. A., Pogodin, Yu. Ya.: On the difference solution of heat conduction equations in curvilinear coordinates (Russian). Dokl. Akad. Nauk SSSR **128**, no. 5 (1959).
15. Yanenko, N. N.: Simple implicit schemes for multidimensional problems (Russian). Proc. of the All-Union Conf. on Comp. Math. and Comp. Techniques, Moscow 1959.
16. Baker jr., G. A., Oliphant, T. A.: An implicit numerical method for solving the two-dimensional heat equation. Quart. Appl. Math. **17**, 361—373 (1960).

17. Baker jr., G. A.: An implicit numerical method for solving the n-dimensional heat equation. Quart. Appl. Math. **17**, 440—443 (1960).
18. Yanenko, N. N.: On the implicit difference computing methods for solving the multidimensional heat conduction equations (Russian). Izv. Vyssh. Uchebn. Zaved., Matematika, no. 4 (23), 148—157 (1961).
19. Oliphant, T. A.: An implicit numerical method for solving two-dimensional time-dependent diffusion problems. Quart. Appl. Math. **19**, 221—229 (1961).
20. Buleev, N. I.: Numerical method for solving the two- and three-dimensional equations of diffusion (Russian). Mat. Sbornik **51** (**93**), no. 2 (1960).
21. D'yakonov, E. G.: Difference schemes with split operators for unsteady equations (Russian). Dokl. Akad. Nauk SSSR **144**, no. 1, 29—32 (1962).
22. D'yakonov, E. G.: Some difference schemes for solving the boundary problems (Russian). Zh. Vych. Mat. **2**, no. 1, 57—79 (1962). English translation: U.S.S.R. Comp. Math. **3**, no. 1, 55—77 (1963).
23. D'yakonov, E. G.: Difference schemes with split operators for multidimensional unsteady problems (Russian). Zh. Vych. Mat. **2**, no. 4, 549—568 (1962). English translation: U.S.S.R. Comp. Math. **3**, no. 4, 581—607 (1963).
24. D'yakonov, E. G.: Difference schemes with split operator for general parabolic equations of second order with variable coefficients (Russian). Zh. Vych. Mat. **4**, no. 2, 278—291 (1964). English translation: U.S.S.R. Comp. Math. **4**, no. 2, 92—110 (1964).
25. Brian, P. L. I.: An infinite difference method of high order of accuracy for the solution of three-dimensional heat conduction problems. A. I. Ch. E. J. **7**, 367—370 (1961).
26. Douglas, J.: Alternating direction method for three space variables. Num. Math. **4**, 41—63 (1962).
27. Saul'ev, V. K.: Integration of equations of parabolic type by the method of networks (Russian). Moscow: Fizmatgiz 1960. English translation New York: Macmillan 1964.
28. Yanenko, N. N.: Some problems of the theory of convergence of difference schemes with constant and variable coefficients (Russian). Proc. of IVth All-Union Math. Conf., vol. 2. Nauka, 1964.
29. Laasonen, P.: Über eine Methode zur Lösung der Wärmeleitungsgleichung. Acta Math. **81**, 309—317 (1949).
30. Landau, L. D., Meiman, N. N., Khalatnikov, I. M.: Numerical integration methods of equations in partial derivatives by the method of networks (Russian). Proc. of IIIrd All-Union Math. Conf., vol. 3. Moscow 1958, pp. 92—100.
31. Ladyzhenskaya, O. A.: Solution of the Cauchy problem for hyperbolic systems by the method of finite differences (Russian). Uchen. Zapis. LGU, Ser. Mat. Nauk **23**, 192—246 (1952).
32. Rouse, C.: A method for the numerical calculation of hydrodynamic flow and radiation diffusion by implicit differencing. J. Soc. Ind. Appl. Math. **9**, no. 1 (1961).
33. Yanenko, N. N., Neuvazhaev, V. E.: A method for the calculation of the gas dynamic motions with non-linear heat conduction (Russian). Trudy MIAN SSSR, no. 74 (1966). English translation in: Difference

methods for solutions of problems of mathematical physics, vol. 1, edited by N. N. Yanenko. Amer. Math. Soc., Providence, R.I., 1967.

34. Godunov, S. K.: Difference methods for the solution of equations of gas dynamics (Russian). Novosibirsk 1962.

35. Yanenko, N. N., Yaushev, I. K.: An absolutely stable integration scheme for the equations of hydrodynamics (Russian). Trudy MIAN SSSR, no. 74 (1966). English translation in: Difference methods for solutions of problems of mathematical physics, vol. 1, edited by N. N. Yanenko. Amer. Math. Soc., Providence, R.I., 1967.

36. Bagrinovskii, K. A., Godunov, S. K.: Difference methods for multidimensional problems (Russian). Dokl. Akad. Nauk SSSR **115**, no. 3, 431—433 (1957).

37. Anuchina, N. N., Yanenko, N. N.: Implicit splitting schemes for hyperbolic equations and systems (Russian). Dokl. Akad. Nauk SSSR **128**, no. 6, 1103—1105 (1959).

38. Godunov, S. K., Zabrodin, A. V.: On difference schemes of the second order of accuracy for multidimensional problems (Russian). Zh. Vych. Mat. **2**, no. 4, 706—708 (1962). English translation: U.S.S.R. Comp. Math. **3**, no. 4, 790—792 (1963).

39. Habetler, G. I., Wachspress, E. L.: Symmetric successive overrelaxation in solving diffusion difference equations. Math. of Comp. **15**, 356—362 (1961).

40. Konovalov, A. N.: The method of fractional steps for the solution of the Cauchy problem for the multidimensional wave equation (Russian). Dokl. Akad. Nauk SSSR **147**, no. 1, 25—27 (1962). English translation: Soviet Math. Dokl. **3**, 1536—1538 (1962).

41. Samarskii, A. A.: Locally one-dimensional difference schemes for multidimensional equations of hyperbolic type in an arbitrary region (Russian). Zh. Vych. Mat. **4**, no. 4, 638—648 (1964). English translation: U.S.S.R. Comp. Math. **4**, no. 4, 21—35 (1964).

42. Friedrichs, K. O.: Symmetric hyperbolic linear differential equations. Comm. Pure Appl. Math. **7**, 345—392 (1954).

43. Anuchina, N. N.: Some difference methods for the systems of hyperbolic equations (Russian). Trudy MIAN SSSR, no. 74 (1966). English translation in: Difference methods for solutions of problems of mathematical physics, vol. 1, edited by N. N. Yanenko. Amer. Math. Soc., Providence, R.I., 1967.

44. Frankel, S.: Convergence rates of iterative treatments of partial differential equations. Math. Tables and Other Aids to Comp. **4**, 65—75 (1950).

45. Young, D.: Iterative methods for solving partial difference equations of elliptic type. Trans. Amer. Math. Soc. **76**, 92—111 (1954).

46. D'yakonov, E. G.: An iterative method for the solution of systems of finite difference equations (Russian). Dokl. Akad. Nauk SSSR **138**, no. 3 (1961).

47. D'yakonov, E. G.: The method of majorizing operator for solving difference analogs of some strongly elliptic systems (Russian). Usp. Mat. Nauk **19**, no. 5 (1964).

48. Il'in, V. P.: Application of the method of alternating directions for the solution of quasilinear equations of parabolic and elliptic types

(Russian). Some Probl. of Appl. and Comp. Math., Collection of Works. Novosibirsk 1966.

49. Enal'skii, V. A.: Motion of particles in an electromagnetic field (Russian). Trudy MIAN SSSR, no. 74 (1966). English translation in: Difference methods for solutions of problems of mathematical physics, vol. 1, edited by N. N. Yanenko. Amer. Math. Soc., Providence, R.I., 1967.
50. Samarskii, A. A.: An economic algorithm in the numerical solution of the system of differential and algebraic equations (Russian). Zh. Vych. Mat. **4**, no. 3, 580—585 (1964). English translation: U.S.S.R. Comp. Math. **4**, no. 3 (1964).
51. Marchuk, G. I., Yanenko, N. N.: Application of the method of fractional steps to the solution of problems of mathematical physics (Russian). Proc. of the All-Union Conf. on Comp. Math., Moscow 1965. — Proc. of the IFIP Congress, New York 1965. — Some Probl. of Appl. and Comp. Math., Collection of Works. Novosibirsk 1966.
52. Richardson, L. D.: The approximate arithmetical solution by finite differences of physical problems involving differential equations, with an application to the stresses in a masonry dam. Phil. Trans. Roy. Soc., London, Ser. A, **210**, 307—317 (1910).
53. Wasow, W. R., Forsythe, G. E.: Finite-difference methods for partial differential equations. New York: Wiley 1960.
54. Markov, V.: On functions with the smallest deviation from zero at a given time interval (Russian). St.-Petersburg 1892.
55. Birkhoff, G., Varga, R. S., Young, D.: Alternating direction implicit method. Advances in computers, vol. 3. New York: Academic Press 1962, pp. 189—273.
56. Birkhoff, G., Varga, R. S.: Implicit alternating direction methods. Trans. Amer. Math. Soc. **92**, 13—24 (1959).
57. Kellog, R. B.: Another alternating direction implicit method. J. Soc. Ind. Appl. Math. **11**, 976 (1963).
58. Timoshenko, S. P., Woinowsky-Krieger, S.: Theory of plates and shells. New York: McGraw-Hill 1959.
59. Konovalov, A. N.: Application of the method of splitting to the dynamic solution of problems of the theory of elasticity (Russian). Zh. Vych. Mat. **4**, no. 4, 760—764 (1964). English translation: U.S.S.R. Comp. Math. **4**, no. 4, 192—198 (1964).
60. Conte, S. D., Dames, R. J.: An alternating direction method for solving the biharmonic equation. Math. Tables and Other Aids to Comp. **12**, no. 63, 198—205 (1958).
61. Konovalov, A. N.: An iterative scheme for solving statistical problems of the theory of elasticity (Russian). Zh. Vych. Mat. **4**, no. 5 (1964). English translation: U.S.S.R. Comp. Math. **4**, no. 5, 217—222 (1964).
62. Douglas jr., J., Gunn, J. E.: Two high order correct difference analogues for the equation of multidimensional heat flow. Math. of Comp. **17**, 71—80 (1963).
63. Samarskii, A. A.: Schemes of higher order of accuracy for multidimensional equations of heat conduction (Russian). Zh. Vych. Mat. **3**, no. 5, 812—840 (1963). English translation: U.S.S.R. Comp. Math. **3**, no. 5, 1107—1146 (1963).
64. Samarskii, A. A., Andreev, V. B.: A difference scheme of higher order of accuracy for equations of elliptic type with several space variables

(Russian). Zh. Vych. Mat. **3**, no. 6, 1006—1013 (1963). English translation: U.S.S.R. Comp. Math. **3**, no. 6, 1373—1382 (1963).

65. Sofronov, I. D.: A difference scheme with diagonal directions of sweeps for solving heat conduction equations (Russian). Zh. Vych. Mat. **5**, no. 2 (1965).

66. Sofronov, I. D.: A difference solution of the equation of heat conduction in curvilinear coordinates (Russian). Zh. Vych. Math. **3**, no. 4, 786—788 (1963). English translation: U.S.S.R. Comp. Math. **3**, no. 4, 1069—1072 (1963).

67. Enal'skii, V. A.: On two systems of higher accuracy for the solution of the Dirichlet problem (Russian). Trudy MIAN SSSR, no. 74 (1966). English translation in: Difference methods for solutions of problems of mathematical physics, vol. 1, edited by N. N. Yanenko. Amer. Math. Soc., Providence, R.I., 1967.

68. Enal'skii, V. A.: An iterative process of higher accuracy (Russian). Proc. of the 3rd Siberian Conf. on Math. and Mech., Tomsk 1964.

69. Marchuk, G. I., Yanenko, N. N.: Solution of a multidimensional kinetic equation by the method of splitting (Russian). Dokl. Akad. Nauk SSSR **157**, no. 6 (1964).

70. Marchuk, G. I., Sultangazin, U. M.: Convergence of the method of splitting for the radiation transfer equations (Russian). Dokl. Akad. Nauk SSSR **161**, no. 1 (1965).

71. Marchuk, G. I., Sultangazin, U. M.: On the solution of the transfer kinetic equation by the method of splitting (Russian). Dokl. Akad. Nauk SSSR **163**, no. 4 (1965).

72. Kantorovich, L. V., Akilov, G. P.: Functional analysis in normalized spaces (Russian). Fizmatgiz, 1959. p. 3

73. Antontsev, S. N., Vasil'ev, O. F., Kuznetsov, B. G., Yanenko, N. N.: Numerical calculation of the spillway (Russian). Some Probl. of Appl. and Comp. Math., Collection of Works. Novosibirsk 1966.

74. Vladimirova, N. N., Kuznetsov, B. G., Yanenko, N. N.: Numerical calculation of the symmetrical flow of viscous incompressible liquid around a plate (Russian). Some Probl. of Appl. and Comp. Math., Collection of Works. Novosibirsk 1966.

75. Douglas jr., J.: The application of stability analysis in the numerical solution of quasi-linear parabolic differential equations. Trans. Amer. Math. Soc. **89**, 484—518 (1958).

76. Godunov, S. K., Semendyaev, K. A.: Difference methods for the numerical solution of problems of gas dynamics (Russian). Zh. Vych. Mat. **2**, no. 1, 3—14 (1962). English translation: U.S.S.R. Comp. Math. **3**, no. 1, 1—12 (1963).

77. Marchuk, G. I.: Numerical algorithm in the solution of the weather forecasting equations (Russian). Dokl. Akad. Nauk SSSR **156**, no. 2 (1964).

78. Marchuk, G. I.: A new approach to numerical solution of the weather forecasting equations. Symposium on Long-range Weather Forecasting, Boulder, U.S.A., 1964.

79. Yanenko, N. N.: On economical implicit schemes (the method of fractional steps) (Russian). Dokl. Akad. Nauk SSSR **134**, no. 5, 1034—1036 (1960). English translation: Soviet Math. Dokl. **1**, 1184—1186 (1961).

80. Yanenko, N. N.: Convergence of the method of splitting for the heat conduction equations with variable coefficients (Russian). Zh. Vych. Mat. **2**, no. 5, 933–937 (1962). English translation: U.S.S.R. Comp. Math. **3**, no. 5, 1094–1100 (1963).

81. Boyarintsev, Yu. E.: Convergence of the method of splitting and the local correctness criterion for difference equations with variable coefficients (Russian). Some Probl. of Appl. and Comp. Math., Collection of Works. Novosibirsk 1966.

82. Lees, M.: Alternating direction methods for hyperbolic differential equations. J. Soc. Ind. Appl. Math. **10**, 610–616 (1962).

83. Lees, M.: Alternating direction and semi-explicit difference methods for parabolic partial differential equations. Num. Math. **3**, 398–412 (1961).

84. Samarskii, A. A.: An economical difference method for the solution of the multidimensional parabolic equation in an arbitrary region (Russian). Zh. Vych. Mat. **2**, no. 5, 787–811 (1962). English translation: U.S.S.R. Comp. Math. **3**, no. 5, 894–926 (1963).

85. Samarskii, A. A.: Convergence of the method of fractional steps for heat conduction equations (Russian). Zh. Vych. Mat. **2**, no. 6, 1117–1121 (1962). English translation: U.S.S.R. Comp. Math. **3**, no. 6, 1347–1354 (1953).

86. Douglas jr., J., Gunn, J. E.: A general formulation of alternating direction methods. Part I. Parabolic and hyperbolic problems. Num. Math. **6**, 428–453 (1964).

87. Faddeev, D. K., Faddeeva, V. N.: Computational methods of linear algebra (Russian). Fizmatgiz, 1960.

88. Yanenko, N. N.: A weak approximation of the systems of differential equations (Russian). Sib. Mat. Zh. **5**, no. 6 (1964).

89. Demidov, G. V., Yanenko, N. N.: The method of weak approximation as constructive method for the solution of the Cauchy problem (Russian). Some Probl. of Appl. and Comp. Math., Collection of Works. Novosibirsk 1966.

90. D'yakonov, E. G.: Some iterative methods for the solution of difference equations which originate during the solution of elliptic equations in partial derivatives by the method of networks (Russian). Comp. Methods and Programming, Collection of Works, vol. 3. MGU, Moscow 1965.

91. Samarskii, A. A.: Difference schemes for multidimensional differential equations of mathematical physics (Russian). Aplikace Matematiky (Praha) **10**, no. 2, 146–163.

92. Kellog, R. B.: An alternating direction method for operator equations. J. Soc. Ind. Appl. Math. **12**, no. 4, 848–854 (1964).

93. Tikhonov, A. N., Samarskii, A. A.: On uniform difference schemes (Russian). Zh. Vych. Mat. **1**, no. 1, 5–63 (1961). English translation: U.S.S.R. Comp. Math. **2**, no. 1, 5–67 (1962).

94. Samarskii, A. A.: Additivity principle for the construction of economic difference schemes (Russian). Dokl. Akad. Nauk SSSR **165**, no. 6, 1253–1256 (1965).

95. D'yakonov, E. G.: Solution of some multidimensional problems of mathematical physics by the method of networks (Russian). Dissertation, MGU, Moscow 1962.

96. Samarskii, A. A.: Economic difference schemes for hyperbolic systems of equations with mixed derivatives and their application to the elasticity equation (Russian). Zh. Vych. Mat. **5**, no. 1, 34–43 (1965). English translation: U.S.S.R. Comp. Math. **5**, no. 1, 44–57 (1965).
97. Marchuk, G. I.: Numerical methods for the solution of the weather forecasting problems and the theory of climate (Russian). Moscow 1965.
98. Gol'din, V. Ya.: A characteristic difference scheme for the unsteady kinetic equation (Russian). Dokl. Akad. Nauk SSSR **133**, no. 4, 748–751 (1960).
99. Yosida, K.: Functional analysis. Berlin-Heidelberg-New York: Springer 1965.
100. Il'in, V. P.: On splitting of difference equations of parabolic and elliptic types (Russian). Sib. Mat. Zh. **6**, no. 6 (1965).
101. D'yakonov, E. G.: Application of difference schemes with split operators in hyperbolic equations with variable coefficients (Russian). Dokl. Akad. Nauk SSSR **151**, no. 4 (1963).
102. Valiullin, A. N., Yanenko, N. N.: Economical difference schemes of higher order accuracy for the polyharmonic equation (Russian). Izv. SO Akad. Nauk SSSR, Ser. Techn. Nauk **3**, no. 13, 88–96 (1967).
103. Fairweather, G., Gourlay, A. R., Mitchell, A. K.: Some high accuracy difference schemes with a splitting operator for equations of parabolic and elliptic type. Num. Math. **10**, 56 (1967).
104. Andreev, V. B.: On the uniform convergence of the difference schemes for the Neumann problem (Russian). Zh. Vych. Mat. **9**, 1285–1298 (1969).
105. Lions, J. L.: Résolution itérative d'inéquations variationnelles par décomposition et éclatement. Séminaire sur les équations aux dérivées partielles, Collège de France, 66–67.
106. Temam, R.: Sur la stabilité et la convergence de la méthode des pas fractionnaires. Ann. Mat. Pura Appl. (IV) **79**, 191–380 (1968).
107. Temam, R.: Une méthode d'approximation de la solution des équations de Navier-Stokes. Bull. Soc. Math. France **96**, 115–159 (1968).
108. Temam, R.: Sur l'approximation de la solution des équations de Navier-Stokes par la méthode des pas fractionnaires (II).
109. Trotter, M. F.: Approximation of semigroups of operators. Pacif. J. Math. **8**, no. 4 (1958).
110. Trotter, M. F.: On products of semigroups of operators. Proc. Amer. Math. Soc. **10**, no. 4 (1959).
111. Chernov, P. R.: Note on product formulas for operation semigroups. J. Funct. Anal. **2** (1968).
112. Yanenko, N. N.: The method of fractional steps for numerical solution of the problems of mechanics of continuous media. Institute of Fundamental Technical Research, Polish Academy of Sciences, Warsaw, Fluid Dynamics Trans. **4**, 135–147 (1969).
113. Yanenko, N. N., Anuchina, N. N., Penenko, V. E., Shokin, Y. I.: On numerical methods for the gas-dynamic flows with great deformations. Numerical Methods in Mechanics of Continua, Inf. Bull., Novosibirsk **1**, no. 1 (1970).

Subject Index

721/32/70—III/18 203